Reactivity and Structure Concepts in Organic Chemistry

Volume 20

Editors:

Klaus Hafner
Charles W. Rees
Barry M. Trost

Jean-Marie Lehn
P. von Ragué Schleyer
Rudolf Zahradník

T. Shono

Electroorganic Chemistry as a New Tool in Organic Synthesis

With 7 Figures and 49 Tables

Springer-Verlag
Berlin Heidelberg New York Tokyo 1984

Professor Tatsuya Shono
Kyoto University
Department of Synthetic Chemistry
Faculty of Engineering
Yoshida, Sakyo
Kyoto 606/Japan

List of Editors

Professor Dr. Klaus Hafner
Institut für Organische Chemie der TH
Petersenstr. 15, D-6100 Darmstadt

Professor Dr. Jean-Marie Lehn
Institut de Chimie, Université de Strasbourg
1, rue Blaise Pascal, B.P. 296/R8, F-67008 Strasbourg-Cedex
Professor Dr. Charles W. Rees, F. R. S. Hofmann
Professor of Organic Chemistry, Department of Chemistry
Imperial College of Science and Technology
South Kensington, London SW72AY, England

Professor Dr. Paul v. Ragué Schleyer
Lehrstuhl für Organische Chemie der Universität Erlangen-Nürnberg
Henkestr. 42, D-8520 Erlangen

Professor Barry M. Trost
Department of Chemistry, The University of Wisconsin
1101 University Avenue, Madison, Wisconsin 53706, U.S.A.

Prof. Dr. Rudolf Zahradník
Tschechoslowakische Akademie der Wissenschaften
J.-Heyrovský-Institut für Physikal. Chemie und Elektrochemie
Máchova 7, 121 38 Praha 2, C.S.S.R.

ISBN 3-540-13070-5 Springer-Verlag Berlin Heidelberg New York Tokyo
ISBN 0-387-13070-5 Springer-Verlag New York Heidelberg Berlin Tokyo

Library of Congress Cataloging in Publication Data

Shono, T. (Tatsuya), 1929– Electroorganic chemistry as a new tool in organic
synthesis. (Reactivity and structure: concepts in organic chemistry; v. 20). Includes
bibliographical references and index.
1. Electrochemistry. 2. Chemistry, Organic — Synthesis. I. Title. II. Series:
Reactivity and structure; vol. 20.
QD273.S56 1984 547'.2 83-20443
ISBN 0-387-13070-5 (U.S.)

Printing: Brüder Hartmann, Berlin. Bookbinding: Lüderitz & Bauer, Berlin
2152/3020-543210

Preface

Although the first electroorganic reaction used in organic synthesis is probably the famous Kolbe electrolysis published in 1849, no other remarkable reactions have been found until the reductive dimerization of acrylonitrile to adiponitrile was developed by Dr. M. M. Baizer of Monsanto Co. in 1964. Since then, the electroorganic chemistry has been studied extensively with the expectation that it is a new useful tool for finding novel reactions in organic synthesis.

The purpose of this book is not to give a comprehensive survey of studies on electrochemical reactions of organic compounds but to show that the electroorganic chemistry is indeed useful in organic synthesis.

Thus, this book has been written under the following policies.

(1) Since this monograph is mainly concerned with organic synthesis, only few studies carried out from the viewpoint of electrochemical, theoretical, or analytical chemistry are mentioned.

(2) Since electroorganic chemistry covers a great variety of reactions, the types of reactions described in this book are selected mainly with regard to their application in organic synthesis. Simple transformations of functional groups are only described in particular cases, and also some well established processes such as the Kolbe electrolysis, pinacolic coupling, and hydrodimerization are only briefly mentioned.

(3) Since many reports have already been published for each type of these reactions, only a limited number of the relevant papers are cited in this book.

(4) In this book, electroorganic reactions are divided into two major types, namely oxidation and reduction Oxidations are further classified according to the structure of the substrate, whereas reductions are classified by the type of reaction.

(5) Since many review articles [1-22] and monographs [23-36] describing electroorganic chemistry from a variety of standpoints have already been published, the author

VI

attempted to present the material under an aspect which is different from these publications.

The author is deeply grateful to Drs. Yoshihiro Matsumura, Shigenori Kashimura, and Kenji Tsubata for their kind assistance in collecting data, and also to Mrs. Yohko Ohmizu for her great efforts in typing the manuscript.

Kyoto, February 1984   Tatsuya Shono

References

1. Weinberg, N. L., Weinberg, H. R.: Chem. Rev. *68*, 449 (1968)
2. Adams, R. N.: Acc. Chem. Res. *2*, 175 (1969)
3. Eberson, L., in: The Chemistry of Carboxylic Acids and Esters (ed.) Patai, S., Chap. 2, New York: Wiley-Interscience 1969
4. Kastening, B.: Chem.-Ing.-Tech. *42*, 190 (1970)
5. Lund, H., in: The Chemistry of the Carbon-Nitrogen Double Bond (ed.) Patai, S., Chap. 11, New York: Wiley-Interscience 1970
6. Lund, H., in: The Chemistry of the Hydroxy Group (ed.) Patai, S., Chap. 5, Vol. I, London: Wiley 1971
7. Wawzonek, S.: Synthesis *1971*, 285
8. Beck, F.: Angew. Chem. *84*, 798 (1972)
9. Lehmkuhl, H.: Synthesis *1973*, 377
10. Eberson, L., Nyberg, K.: Acc. Chem. Res. *6*, 106 (1973)
11. Casanova, J., Eberson, L., in: The Chemistry of the Carbon-Halogen Bond (ed.) Patai, S., Chap. 15, Vol. 2, New York: Wiley 1973
12. Chambers, J. Q., in: The Chemistry of Quinonoid Compounds (ed.) Patai, S., Chap. 14, Part 2, New York: Wiley 1974
13. Thomas, F. G., Boto, K. G., in: The Chemistry of Hydrazo, Aza and Azoxy Groups (ed.) Patai, S., Chap. 12, Vol. 1, New York: Wiley 1975
14. Lund, H., in: The Chemistry of Amidines and Imidates (ed.) Patai, S., Chap. 5, New York: Wiley 1975
15. Mairanovsky, V. G.: Angew. Chem. *88*, 283 (1976)
16. Eberson, L., Nyberg, K.: Tetrahedron *32*, 2185 (1976)
17. Fry, A. J., Reed, R. G., in: The Chemistry of Double-Bonded Functional Groups (ed.) Patai, S., Chap. 5, Vol. 1, New York: Wiley 1977
18. Hammerich, O., Parker, V. D., in: The Chemistry of Cyanates and Their Thioderivatives (ed.) Patai, S., Chap. 9, Vol. 1, New York: Wiley 1977
19. Utley, J. H. P., Lines, R., in: The Chemistry of the Carbon-Carbon-Triple Bond (ed.) Patai, S., Chap. 17, Vol. 2, New York: Wiley-Interscience 1978
20. Fry, A. J., in: The Chemistry of the Diazo and Diazonium Group (ed.) Patai, S., Chap. 10, Vol. 1, New York: Wiley-Interscience 1978

21. Shono, T., in: The Chemistry of Ethers, Crown Ethers, Hydroxyl Groups and Their Sulphur Analogues (ed.) Patai, S., Chap. 8, Vol. 1, New York: Wiley-Interscience 1980
22. Schäfer, H. J.: Angew. Chem. Int. Ed. Engl. *20*, 911 (1981)
23. Mann, C. K., Barnes, K. K.: Electrochemical Reactions in Nonaqueous Systems, New York: Dekker 1970
24. Eberson, L., Schäfer, H. J.: Organic Electrochemistry, Berlin, Heidelberg, New York: Springer 1971
25. Yeager, E., Salkind, A. J.: Techniques in Electrochemisty, New York: Wiley 1972
26. Fry, A. J., Dryhurst, G.: Organic Electrochemistry, Berlin– Heidelberg–New York: Springer 1972
27. Fry, A. J.: Synthetic Organic Electrochemistry, New York: Harper & Row 1972
28. Tomilov, A. P. et al.: The Electrochemistry of Organic Compounds, New York: Halsted Press 1972
29. Baizer, M. M., Lund, H. (ed.): Organic Electrochemistry, New York: Dekker 1983
30. Weinberg, N. L. (ed.): Techniques of Electroorganic Synthesis, Part I—III, New York: Wiley-Interscience 1974, 1975, 1982
31. Rifi, M. R., Covitz, F. H.: Introduction to Organic Electrochemistry, New York: Dekker 1974
32. Beck, F.: Electroorganische Chemie, Weinheim: Verlag Chemie 1974
33. Sawyer, D. T., Roberts, J. L., Jr.: Experimental Electrochemistry for Chemists, New York: Wiley-Interscience 1974
34. Ross, S. D., Finkelstein, M., Rudd, E. J.: Anodic Oxidation, New York: Academic Press 1975
35. Bard, A. J., Lund, H.: Encyclopedia of Electrochemistry of the Elements, Vols. 11–13, New York: Dekker 1978, 1979
36. Swann, S., Jr., Alkire, R. C.: Bibliography of Electroorganic Syntheses, 1801–1975, Princeton, N.J.: The Electrochemical Society 1980

Table of Contents

X

1. Introduction

The term electroorganic chemistry is often used in the same meaning as organic electrochemistry. Although the difference between these two terms is not always clear, the latter seems to emphasize electrochemistry rather than organic chemistry and to involve electrochemical, electrotheoretical, and electroanalytical studies of organic compounds. On the other hand, the former is basically organic chemistry and offers variety of potentials which are beneficial to organic synthesis as briefly surveyed below.

1.1 Inversion of Polarity of Substrates

In electroorganic reactions, the active species is generated on the electrode surface through electron transfer between a substrate molecule and the electrode as shown in Eq. (1),[1] in which the substrate molecule is transformed to a cation radical or an anion radical, depending on the direction of electron transfer. When the substrate molecule is a radical or ionic species, the transformation of the substrate is such as shown in Eq. (2).

$$A^{2-} \underset{+e}{\overset{-e}{\rightleftharpoons}} A^{\cdot -} \underset{+e}{\overset{-e}{\rightleftharpoons}} A \underset{+e}{\overset{-e}{\rightleftharpoons}} A^{\cdot +} \underset{+e}{\overset{-e}{\rightleftharpoons}} A^{2+} \tag{1}$$

$$A^{-} \underset{+e}{\overset{-e}{\rightleftharpoons}} A^{\cdot} \underset{+e}{\overset{-e}{\rightleftharpoons}} A^{+} \tag{2}$$

As is well known, an organic reaction between two substrate molecules is not achievable when the polarity of the reaction site is the same in both substrates. Thus, the reaction usually takes place between a nucleophilic site (Nu) and an electrophilic site (E). In organic synthesis, however, it is not uncommon that reaction between two groups of the same polarity is required to synthesize the target compound. Then, inversion of the polarity of one of the groups is the necessary means of achieving such a type of reaction, though this inversion (Umpolung) is not always facile in organic chemistry. As Eqs. (1) and (2) clearly show, in an electroorganic reaction, the generation

---

[1] —e, removal of one electron; —2e, removal of two electrons; —[e], removal of electrons

1

of active species through electron transfer between a substrate and electrode always involves inversion of polarity of the substrate. Thus, this facile inversion of polarity makes electroorganic chemistry a unique tool in organic synthesis.

1.2 Interface Reactions

As described above the active species is formed on the electrode surface which is an interface between solid and solution. The active species formed under such special circumstances shows unique characteristics in reactivity.

1.2.1 Stereochemistry

One of these characteristics is stereospecificity. The stereospecificities observed in the acetoxylation of methylcyclohexenes (Section 2.2.2.) and in the intramolecular cyclization of non-conjugated olefinic ketones (Section 3.1.1.) are typical examples of stereospecific reactions taking place at the interface. Another interesting example of the stereospecificity has been observed in the anodic acetoxylation of some cyclic dienol acetates carried out in acetic acid containing potassium acetate as the supporting electrolyte (3).

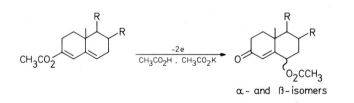

(3)

The same products are also obtained by oxidation of the dienol acetates with perbenzoic acid or by oxidation of the corresponding enones with the liver microsomal oxidation system (4, 5).

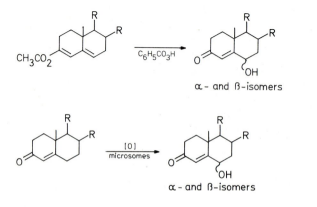

(4)

(5)

The stereoconfiguration of the products obtained by the anodic method shows a remarkable similarity with the microsomal oxidation products whereas perbenzoic acid oxidation in a homogeneous solution exhibits poor stereospecificity (Table 1) [1].

Table 1. Stereospecificity in Anodic, Chemical, and Microsomal Oxidations

Substrate	Product ratio (β-isomer/α-isomer)		
	Anodic method	Microsomal method	Chemical method
	13.9	14.1	3.0
	12.3	11.8	2.8
	10.3	8.4	2.9
	13.3	12.1	3.8

The similarity of anodic and microsomal oxidations may be explained by the fact that both types of oxidations take place at the interface.

1.2.2 Distribution of Active Species

The active species generated on the electrode surface generally reacts with other reagents before it is diffused uniformly into the solution, whereas in the usual organic homogeneous reaction the distribution of active species is uniform in solution. Due to this difference the electrogenerated active species displays unique characteristics. For example, as shown in Section 3.1.2., the carbanion formed by the reduction of an iminium cation is alkylated with alkyl halides in high yields under highly acidic reaction conditions (6).

$$-CH=\overset{+}{N}R \xrightarrow{+2e} -\overset{-}{C}H-NHR \xrightarrow{R'X} -CH-NHR \qquad (6)$$
$$\quad\;\; | \qquad\qquad\qquad\qquad\qquad\qquad\quad\; |$$
$$\quad\;\; H \qquad\qquad\qquad\qquad\qquad\qquad\quad R'$$

3

1. Introduction

Since it is not possible to alkylate the carbanion in acidic homogeneous solution before it is protonated, reaction (6) is assumed to involve alkylation of the carbanion in the vicinity of the electrode before it is diffused into the acidic solution.

The famous Kolbe electrolysis seems to be a typical example showing the uniqueness of the distribution of the electrogenerated active species. Thus, the free-radical species formed at rather high concentration on the anode surface through anodic oxidation of a carboxylate anion dimerizes before it is diffused into solution, whereas the same radical species generated in a homogeneous solution by the usual chemical method forms a dimer as a minor product but mainly abstracts hydrogen from hydrogen donors, e.g. from the solvent.

Reference

1. Shono, T., Toda, T., Oshino, N.: Tetrahedron Lett. *25*, 91 (1984)

2. Anodic Oxidations

2.1 Anodic Cleavage of Aliphatic Carbon—Hydrogen Bonds and Carbon—Carbon Single Bonds

Direct anodic oxidation of alkanes may be performed if they have ionization potentials lower than about 10 eV [1][1]. Such oxidations can be classified into two types of reactions, cleavage of C—H bonds (1) and C—C bonds (2).

$$RH \xrightarrow{-2e} R^+ + H^+ \tag{1}$$

$$R-R \xrightarrow{-2e} 2R^+ \tag{2}$$

The extremely high oxidation potentials of alkanes, however, make it difficult to carry out the oxidation in commonly used solvents like acetonitrile. Since the first intermediates generated in these oxidations are carbonium ions as illustrated by Eqs. (1) and (2), their stabilization with strongly acidic solvents like anhydrous fluorosulfonic acid lowers the oxidation potentials of the hydrocarbons [2][2].

2.1.1 Cleavage of Carbon—Hydrogen Bonds

The controlled potential electrolysis of cyclohexane carried out at 1.85 V in fluorosulfonic acid containing 1.15 M acetic acid yields an α,β-unsaturated ketone as a single product in 30% current yield (3).

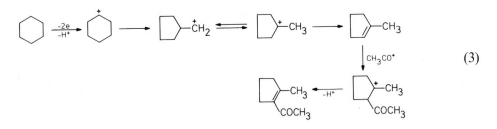

$$\tag{3}$$

---

[1] The ionization potential and half-wave oxidation potential ($E_{1/2}$, 0.14 M $(C_2H_5)_4$-NBF_4/CH_3CN) of some alkanes are as follows: isopentane, 10.10 eV (3.00 V); 2-methylpentane, 10.00 (3.01 V); 2,2-dimethylbutane, 10.19 (3.28 V); octane, 10.24 (>3.4 V).
[2] The oxidation potentials of some alkanes in fluorosulfonic acid containing 1.15 M acetic acid are as follows: isopentane, 1.8 V *vs.* Pd/H_2; 2-methylpentane, 1.68 V; octane, 1.64 V.

The acetyl cation is formed from acetic acid. Although α,β-unsaturated keto-nes are always formed in the anodic oxidation of alkanes in fluorosulfonic acid containing acetic acid, the products are mixtures of isomers. This anodic oxidation is therefore not widely utilized in organic synthesis. In more weakly acidic solution, different products are obtained [3]. Thus, the anodic oxidation of cyclohexane in CH_2Cl_2 containing 2.5 M CF_3CO_2H yields cyclohexyl trifluoroacetate $(Y = 91\%)$ and cyclohexyl chloride $(Y = 4\%)$.

$$
\text{(cyclohexane)} + CF_3CO_2H \xrightarrow[CH_2Cl_2]{-2e} \text{(cyclohexyl-OCOCF}_3) + \text{(cyclohexyl-Cl)}
$$
(4)

The oxidation of alkanes like decane gives, however, mixtures of products.

In the anodic oxidation, adamantane is a unique compound among alkanes. It has a rather low oxidation potential, and its anodic oxidation in acetonitrile affords acetoamidoadamantane in 90% yield [4][3].

$$
\text{(adamantane)} \xrightarrow[CH_3CN, H_2O]{-2e} \text{(adamantyl-NHCOCH}_3)
$$
(5)

This reaction involves direct oxidation of adamantane to adamantyl cation as the active intermediate. A variety of substituted adamantanes can be oxidized forming similar products in satisfactory yields.

2.1.2 Cleavage of Carbon—Carbon Single Bonds

The direct anodic cleavage of saturated aliphatic carbon-carbon bonds is only possible if an electron is removed from the highest occupied molecular orbital (HOMO) of the C—C bond e.g. due to the presence of strain in the bond [5]. The oxidation potentials ($E_{1/2}$ V vs. SCE) of some strained hydrocarbons are shown below together with the ionization potentials (IP eV) [5–6].

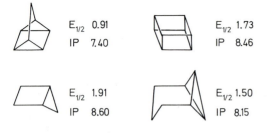

$E_{1/2}$ 0.91   IP 7.40

$E_{1/2}$ 1.73   IP 8.46

$E_{1/2}$ 1.91   IP 8.60

$E_{1/2}$ 1.50   IP 8.15

---

[3] Oxidation potential of adamantane: $E_p/2$, 2.36 V, sweep rate 0.1 V/s, 10^{-2} M substrate, 10^{-1} M $(C_4H_9)_4NBF_4/CH_3CN$, 10^{-1} M Ag/Ag^+ reference.

Tetramethylcyclopropane is the simplest strained hydrocarbon which is easily oxidized by the anodic method in methanol to give two products with a total yield of 71 % [6]:

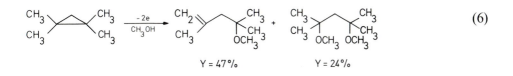

$$Y = 47\% \qquad\qquad Y = 24\%$$

(6)

The cleavage of the C—C bond shows a remarkable selectivity since the most substituted bond is split, and products resulting from cleavage of the other bonds are not formed.

Furthermore, the anodic cleavage is in sharp contrast to the selectivity of C—C bond cleavage observed in the reaction of tetramethylcyclopropane with acidic methanol in which the less substituted bond is cleaved.

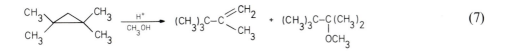

(7)

This difference of selectivity may be explained as follows: In the anodic oxidation, the electron is transferred to the anode from the most strained C—C bond where the ionization potential is lowest whereas in the acidic solvolysis, the proton attacks the less hindered site.

The marked difference between the reaction pattern of the anodic oxidation and that of the common chemical oxidation is also demonstrated by the following reactions [7].

The anodic C—C bond cleavage of bicyclo[4.1.0]heptane in methanol takes place exclusively at the internal bond which is the most strained (8).

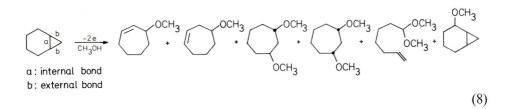

a: internal bond
b: external bond

(8)

On the other hand, the selectivities (external/internal) observed in the bond cleavage of the same compound by acidic solvolysis and by oxidation with metallic oxidizing agents are lower, and the cleavage mainly occurs at the external bond (Table 1).

Similar exclusive external bond cleavage is observed in the acidic solvolysis of cis- and trans-bicyclo[6.1.0]nonanes and cis-bicyclo[5.1.0]octane [8]. In

Table 1. Selectivity of Bond Cleavage

Reaction conditions	Anodic cleavage	TsOH/CH$_3$OH 24 h reflux	TsOH/CH$_3$CO$_2$H 24 h 47 °C	H$_2$SO$_4$/CH$_3$CO$_2$H 41 h 46.5 °C	Pb(OAc)$_4$	Tl(OAc)$_3$
Selectivity (external/internal)	0/100	87.2/12.8	88.9/11.1	79.2/20.8	71/29	91/9

acidic solvolysis, cleavage of the internal bond occurs only in the reaction of highly strained compounds such as bicyclo[2.1.0]pentane [9] and *trans*-bicyclo[5.1.0]octane. On the other hand, the anodic C—C bond cleavage of bicyclo[4.1.0]heptane and bicyclo[3.1.0]hexane is completely internal.

The anodic oxidation of tricyclo[4.1.0.0^{2,7}]heptane in methanol also involves internal bond cleavage (9) [5].

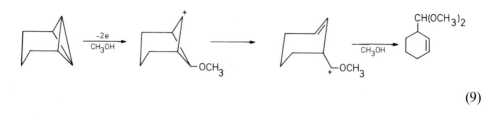

$$(9)$$

References

1. Fleischmann, M., Pletcher, D.: Tetrahedron Lett. *1968*, 6255
2. Bertram, J., Fleischmann, M., Pletcher, D.: Tetrahedron Lett. *1971*, 349
3. Fritz, H. P., Würminghausen, T.: J. Chem. Soc., Perkin Trans. (1) *1976*, 610
4. Koch, V. R., Miller, L. L.: J. Am. Chem. Soc. *95*, 8631 (1973)
5. Gassman, P. G., Yamaguchi, R.: J. Am. Chem. Soc. *101*, 1308 (1979)
6. Shono, T., Matsumura, Y.: Bull. Chem. Soc. Jpn. *48*, 2861 (1975)
7. Shono, T., Matsumura, Y., Nakagawa, Y.: J. Org. Chem. *36*, 1771 (1971)
8. Wiberg, K. B., de Meijere, A.: Tetrahedron Lett. *1969*, 519
9. Lalonde, R. T., Forney, L. S.: J. Am. Chem. Soc. *85*, 3767 (1963)

2.2 Oxidation of Carbon—Carbon Double Bonds

2.2.1 Introduction

As the oxidation potentials of simple olefins (e.g. (C$_2$H$_5$)$_2$C=CH$_2$, E$_{1/2}$ = 2.17 V *vs.* SCE) clearly show, carbon-carbon double bonds are usually anodically oxidized unless electron-withdrawing groups located on the unsaturated bonds attract electrons from the unsaturated systems to shift the oxidation potentials beyond those accessible by anodic oxidation.

On the other hand, electron-donating groups on the unsaturated bonds facilitate oxidation of the latter.

The initiation step of the anodic oxidation involves removal of an electron from the unsaturated bonds leading to a cation radical as the first reactive intermediate. Depending on the structure of the unsaturated compounds, a variety of reactions will take place after the formation of the first intermediate. Thus, typical reactions are addition of nucleophiles (1), allylic substitution (2), and dimerization (3).

$$
\overset{|}{\underset{|}{C}}=\overset{|}{\underset{|}{C} } \xrightarrow{-e} \overset{|}{\underset{|}{C^{+}}}-\overset{|}{\underset{|}{\overset{\cdot}{C}}} \xrightarrow[2\,Nu]{-e} Nu-\overset{|}{\underset{|}{C}}-\overset{|}{\underset{|}{C}}-Nu \tag{1}
$$

$$
-\overset{|}{\underset{|}{C}}-\overset{|}{\underset{|}{C}}=\overset{|}{\underset{|}{C}} \xrightarrow{-e} -\overset{|}{\underset{|}{C}}-\overset{|}{\underset{|}{\overset{+}{C}}}-\overset{\cdot}{\underset{|}{C}} \xrightarrow[Nu]{-e} -\underset{\underset{Nu}{|}}{\overset{|}{C}}-\overset{|}{C}=\overset{|}{\underset{|}{C}} \tag{2}
$$

$$
2\overset{|}{\underset{|}{C}}=\overset{|}{\underset{|}{C}} \xrightarrow{-2e} 2\overset{|}{\underset{|}{\overset{+}{C}}}-\overset{\cdot}{\underset{|}{C}} \xrightarrow[2\,Nu]{} Nu-\overset{|}{\underset{|}{C}}-\overset{|}{\underset{|}{C}}-\overset{|}{\underset{|}{C}}-\overset{|}{\underset{|}{C}}-Nu \tag{3}
$$

The mechanism of each of these reactions is however more complicated than that shown in the above equations.

2.2.2 Oxidation of Simple Olefins

Although the oxidation potentials of monoalkylated olefins are in a region hardly accessible by the electrochemical method, most of di-, tri-, and tetraalkylated olefins may be oxidized by the anodic technique. Oxidation potentials of some olefins are compiled in Table 1 [1, 2].

Table 1.[a] Oxidation Potentials of Olefins

Olefin	Oxidation potential $E_{1/2}$ (V vs. SCE)	Olefin	Oxidation potential $E_{1/2}$ (V vs. SCE)
1-octene	2.8	cyclopentene	1.96
2-octene	2.3	β-pinene	1.89
1,1-diethylethylene	2.17	1-methylcyclohexene	1.70
cyclohexene	2.14	tetramethylethylene	1.48
norbornene	2.02	α-pinene	1.41

[a] Supporting electrolyte, $LiClO_4$;   solvent, CH_3CN

2. Anodic Oxidations

In general, the anodic oxidation of simple olefins in nucleophilic solvents yields products resulting from both allylic substitution and oxidative addition of nucleophiles. Cyclohexene has been studied extensively as the starting compound [3a–e]. The anodic oxidation of cyclohexene in methanol or acetic acid using tetraethylammonium p-toluenesulfonate as a supporting electrolyte gives three types of products, that is allylic substitution, oxidative addition, and rearrangement products (4).

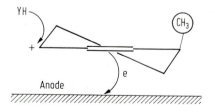

$$Y = OAc, OCH_3$$

(4)

An interesting stereospecificity has been observed in the allylically substituted products obtained from methylcyclohexenes.

As shown in Table 2, anodically acetoxylated methylcyclohexene shows a remarkable dominance of the cis-isomer whereas the Kharash-Sosnovsky reaction yields mainly the thermodynamically stable trans-isomer.

Table 2. Stereochemistry of Acetoxylation

Methylcyclohexene	Acetoxylated compound	cis/trans Ratio in anodic reaction	Kharasch-Sosnovsky reaction
CH₃ ◯	CH₃ ◯ OAc	2.83	0.23
CH₃ ◯	CH₃ ◯ OAc	3.41	0.29

This result may be explained by the stereoselective adsorption of methylcyclohexene on the anode as shown in Fig. 1.

It is likely that methylcyclohexene is adsorbed on the anode facing the methyl group to the side of the solution. After the cationic intermediate

Fig. 1

is formed on the anode, the nucleophilic solvent YH attacks the intermediate from the same side of methyl group.

The mechanism of oxidation of cyclohexene has been shown to involve direct removal of one electron from the double bond to generate a cation radical intermediate (5) [2].

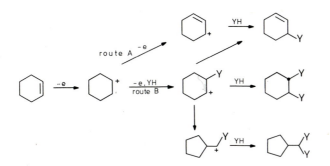

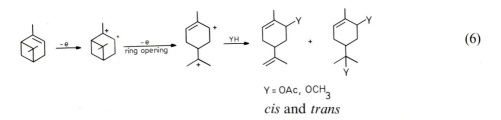

(5)

The relative ratio of routes A and B may be controlled by the nucleophilicity of YH. The intermediary formation of cationic species has been supported by a variety of evidences. One of these evidences involves comparison of the product distribution in the anodic allylic substitution of methylcyclohexenes with that obtained by the Kharasch-Sosnovsky reaction of the same compounds. The Wagner-Meerwein type rearrangement observed in the anodic oxidation of α- and β-pinenes also strongly suggests the presence of cationic intermediates.

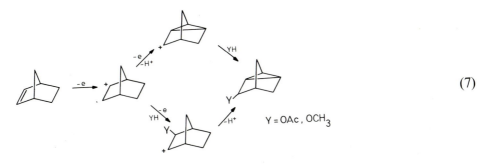

(6)

$$Y = OAc, OCH_3$$
cis and *trans*

Furthermore, the formation of nortricyclenes from norbornenes may also be explained to proceed via intermediate cationic species [4].

(7)

$$Y = OAc, OCH_3$$

2. Anodic Oxidations

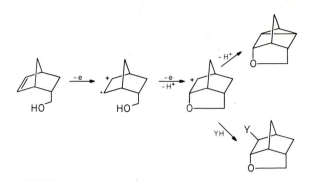

(8)

When the olefin is the most easily oxidizable compound in the reaction system, the reaction mainly proceeds according to the mechanism involving direct one-electron oxidation of the olefin as shown in Eq. (5). If the oxidation potential of the supporting electrolyte or solvent is lower than or comparable to that of olefin, a mechanism involving the participation of radicals or cationic species formed by the oxidation of the supporting electrolyte or solvent is also possible.

$$S^- \xrightarrow{-e} S\cdot \xrightarrow{-e} S^+$$

S^- : anionic part of solvent or supporting electrolyte

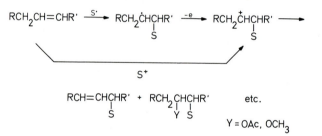

(9)

In fact, when tetraethylammonium chloride or bromide is used as the supporting electrolyte, the anodic oxidation of cyclohexene in acetic acid yields 1-halo-2-acetoxycyclohexane along with the allylically substituted product, the yield of addition product being excellent when X is Br[4].

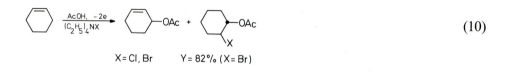

(10)

---

[4] The reaction induced by the oxidation of the halide ion will be described in Section 2.5.

12

The extent of the participation of the mechanism (9) to the entire reaction depends on the relative value of the oxidation potentials of supporting electrolyte, solvent and olefin, and also on the lifetime and reactivity of S· or S⁺. In an exceptional case where the radical species (S·) generated from the solvent or electrolyte possesses a sufficiently long lifetime, the reaction mechanism involving radical coupling between the allylic radical and S· may be possible.

The yield of this anodic allylic substitution may be improved by modification of the reaction conditions. For instance, when the reaction is carried out in the presence of a cation exchange resin (Dowex 50W-X8) a higher yield of acetaminocyclohexene is obtained [5].

Anodic allylic substitution is applicable to a long-chain alkenes possessing a higher oxidation potential [6]. Oxidative addition to olefins involving a chain reaction has been used in the synthesis of adamantylidene-adamantane dioxetane [7].

2.2.3 Oxidation of Conjugated and Nonconjugated Dienes

Conjugated dienes are generally more susceptible to oxidation than simple olefins (Table 3) [8].

Table 3. Peak Potentials of Oxidation of Dienes[a]

Diene	E_p (V vs. Ag/Ag⁺)	Diene	E_p (V vs. Ag/Ag⁺)
	2.0		1.55, 1.70
	1.75		1.48
	1.50		1.36

[a]Glassy carbon; solvent, CH_3OH; supporting electrolyte, 0.5 M $NaClO_4$

When using a carbon electrode, the anodic oxidation of conjugated dienes such as isoprene, piperylene, cyclopentadiene, and 1,3-cyclohexadiene in methanol or acetic acid containing tetraethylammonium p-toluenesulfonate as the supporting electrolyte gives mainly oxidative 1,4-addition products (11).

(11)

2. Anodic Oxidations

Some results are listed in Table 4 [9]; 1,3-cyclooctadiene yields a considerable amount of the allylically substituted product.

The oxidation of conjugated dienes has been successfully applied to the synthesis of 1,4-diacetoxy-2-cyclopentenes (Table 5) and allethrolone (12) [10].

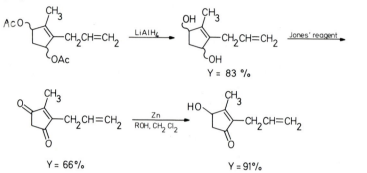

(12)

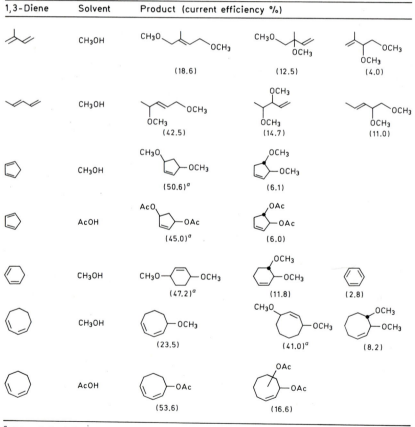

Table 4. Oxidation of Conjugated Dienes

1,3-Diene	Solvent	Product (current efficiency %)		
![diene]	CH₃OH	CH₃O~~~OCH₃ (18.6)	CH₃O~~~OCH₃ (12.5)	~~~OCH₃ OCH₃ (4.0)
![diene]	CH₃OH	~~~OCH₃ OCH₃ (42.5)	OCH₃ ~~~OCH₃ (14.7)	~~~OCH₃ OCH₃ (11.0)
![cyclopentadiene]	CH₃OH	CH₃O-[ring]-OCH₃ (50.6)ᵃ	[ring]-OCH₃ OCH₃ (6.1)	
![cyclopentadiene]	AcOH	AcO-[ring]-OAc (45.0)ᵃ	[ring]-OAc OAc (6.0)	
![cyclohexadiene]	CH₃OH	CH₃O-[ring]-OCH₃ (47.2)ᵃ	[ring]-OCH₃ OCH₃ (11.8)	[ring] (2.8)
![cyclooctadiene]	CH₃OH	[ring]-OCH₃ (23.5)	CH₃O-[ring]-OCH₃ (41.0)ᵃ	[ring]-OCH₃ OCH₃ (8.2)
![cyclooctadiene]	AcOH	[ring]-OAc (53.6)	[ring]-OAc OAc (16.6)	

ᵃMixture (1:1) of cis and trans isomers

14

Table 5. Synthesis of 1,4-Diacetoxy-2-cyclopentenes

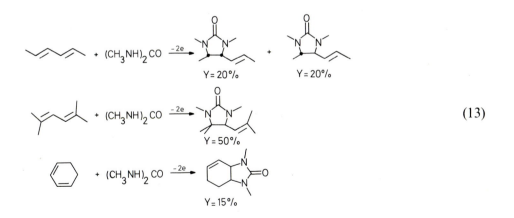

R^1	R^2	Current efficiency (2 F/mol)
CH_3	CH_3	43
CH_3	i-Pr	61
CH_3	$CH_2=CH-CH_2-$	46
CH_3	Pr	57
H	$CH_2=CH-CH_2-$	40
H	i-Pr	57
H	H	45

The oxidative 1,4-addition of nucleophiles to conjugated dienes seems to possess a high potentiality in organic synthesis, though further investigations are desirable.

The formation of a variety of 2-imidazolidinones by the anodic oxidation of conjugated dienes in the presence of N,N'-dimethyl urea can be utilized to introduce amino functional group into dienes, though yields are so far not always satisfactory [11].

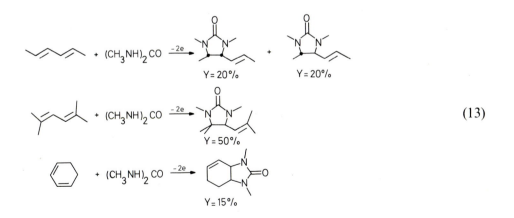

(13)

Using a graphite or carbon-cloth anode, the oxidation of conjugated dienes in 0.5 M methanolic solution of $NaClO_4$ mainly yields dimerized products along with small amounts of monomeric and trimeric compounds [8].

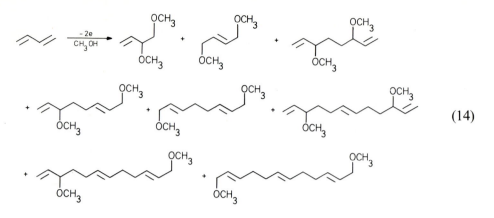

(14)

The product distribution is highly controlled by the anodic material. The use of platinum or glassy carbon mainly gives monomeric products. Other dienes such as isoprene, 1,3-cyclohexadiene, 2,4-hexadiene, 1,3-pentadiene, and 2,3-dimethyl-1,3-butadiene yield dimeric products, though the products are generally complex mixtures of isomers of monomeric, dimeric, and trimeric compounds. The control of the product distribution is the most important problem in the application of electrodimerization of conjugated dienes to organic synthesis.

Compared with simple aliphatic olefins and conjugated dienes, the behavior of nonconjugated dienes in anodic oxidation is unique [12]. The possible reaction pathway of the oxidation of nonconjugated dienes can be classified into two categories (15).

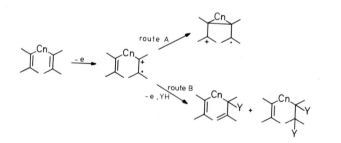

(15)

In route A, one electron is removed from one double bond to generate a cation radical followed by a transannular reaction of the cation radical with the other double bond to form a new carbon-carbon bond. On the other hand, in route B, allylic substitution or oxidative addition at one double bond takes place without intramolecular interaction between both double bonds. As exemplified by the anodic oxidation of 4-vinylcyclo-hexene in methanol (16), such dienes as 4-vinylcyclohexene, limonene, and 1,5-cyclooctadiene yield only products via route B.

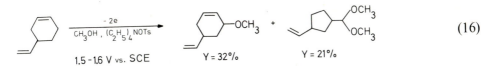

$$\text{(16)}$$

Y = 32%   Y = 21%

On the other hand, the electrooxidation of norbornadiene or bicyclo-[2.2.2]octa-2,5-diene, in which two double bonds are suitably arranged for the transannular interaction to take place, yields products via route A.

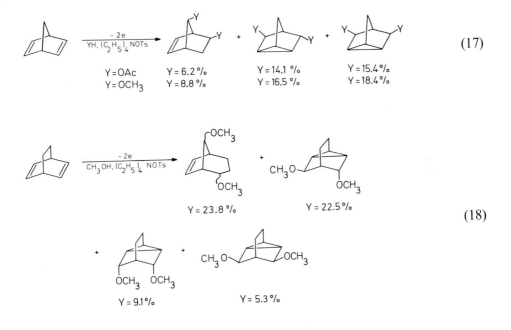

$$\text{(17)}$$

$$\text{(18)}$$

The difference between dienes reacting according to route A and those according to route B is clearly reflected in their oxidation potentials (Table 6).

Thus, the oxidation potentials of the latter types of dienes are substantially the same as those of the corresponding monoolefins whereas norbornadiene and bicyclo[2.2.2]octadiene, which react via route A, show much lower oxidation potentials than those of norbornene and cyclohexene, respectively.

This result suggests that in the anodic oxidation of dienes proceeding via route A, the cation radical formed from one double bond is stabilized through transannular interaction with another double bond. The oxidation potentials of 2-substituted norbornadiene (A), 2-substituted bicyclo[2.2.2]-octa-2,5-dienes (B), and 4-substituted [2.2]paracyclophanes (C) clearly show that the transannular interaction between the two double bonds contributes already at the stage of the first electron transfer (Table 7, Fig. 2). Thus, in the compounds A, B, and C, the electron is removed from the unsaturated bond which does not bear the electron-withdrawing group, while the substituents

17

2. Anodic Oxidations

Table 6. Oxidation Potentials of Dienes and the Corresponding Olefins (V *vs.* SCE)

Substrate	Oxidation potential	Substrate	Oxidation potential
1-methylcyclohexene	1.70	cyclohexene	2.14
limonene	1.67	norbornene	2.02
cyclooctene	2.01	norbornadiene	1.54
1,5-cyclooctadiene	2.07	bicyclo [2.2.2] octadiene	1.82

Solvent, CH_3CN;   supporting electrolyte, 0.1M $LiClO_4$

strongly affect the oxidation potentials, indicating transannular interaction at the stage of electron transfer.

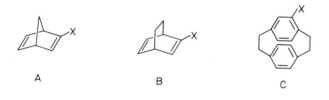

A   B   C

Table 7. Oxidation Potential of A, B, and C

Substituent (X)	Oxidation potential (V *vs.* SCE)		
	A	B	C
H	1.54	1.82	1.47
CO_2CH_3			1.61
$CO_2C_2H_5$	1.85	2.11	
$COCH_3$	1.85	2.07	1.57
CN	1.99	2.22	1.65
NO_2			1.72

The electrooxidation of allenes in methanol proceeds stepwise yielding a tetramethoxylated compound as one of the main products [13].

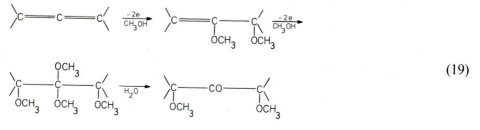

(19)

18

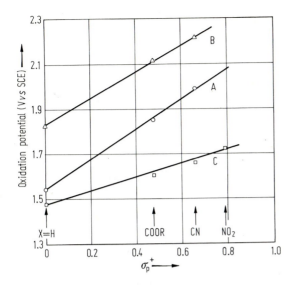

Fig. 2. Plots of the Oxidation Potentials of A, B, and C vs. σ_p^+

2.2.4 Oxidation of Arylolefins

Although the oxidative addition of nucleophiles to the double bond of arylolefins has well been known [14], the most interesting reaction from the synthetic point of view is oxidative dimerization. Using a graphite electrode, the anodic oxidation of styrene in methanol containing $NaOCH_3$ and $NaClO_4$ as supporting electrolytes yields 1,4-dimethoxy-1,4-diphenylbutane in 64% yield [15].

When $NaOCH_3$ and sodium camphorsulfonate are used as the supporting electrolyte instead of $NaOCH_3$ and $NaClO_4$, the same reaction affords *trans,trans*-1,4-diphenylbutadiene (36%) and 1-methoxy-1,4-diphenyl-3-bu-tene (28%).

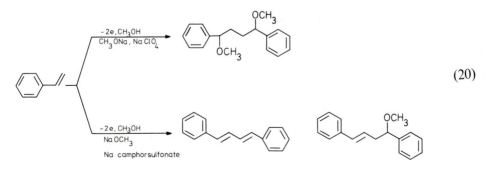

(20)

19

2. Anodic Oxidations

The oxidation of α-methylstyrene at 1.6–1.7 V ($vs.$ Ag/Ag$^+$) using NaOCH$_3$ and NaClO$_4$ as supporting electrolytes yields (E,E)-2,5-diphenyl-2,4-hexadiene in 67% yield.

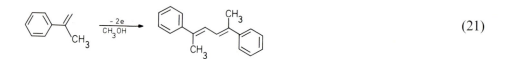

$$(21)$$

The pattern of the reaction strongly depends on the reaction conditions and the structure of the starting compounds. Indene is converted into 1,2-dimethoxyindane (45%) and 1,1′-dimethoxy-2,2′-bisindane (26%) [15, 16].

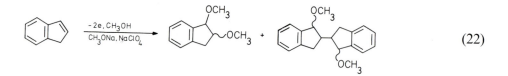

$$(22)$$

The electrooxidative dimerization of styrene in aqueous acetonitrile gives a variety of products with 2,5-diphenyltetrahydrofuran as the main product (40–50%) [17].

$$(23)$$

The mechanism of the dimerization of arylolefins is complex and controlled by a variety of factors. 4,4′-Dimethoxystilbene (DMOS) has been known to be a suitable model compound for analyzing in detail the mechanism of dimerization [18]. In the absence of nucleophiles, the dimerization of DMOS has been proposed to proceed according to two different mechanisms (radical and ECE route):

ECE route

DMOS $\rightleftharpoons$ DMOS$^{+\cdot}$ + e
DMOS$^{+\cdot}$ + DMOS $\xrightarrow{K1}$ (DMOS)$_2^{+\cdot}$
(DMOS)$_2^{+\cdot}$ $\rightleftharpoons$ (DMOS)$_2^{2+}$ + e

radical route

DMOS $\rightleftharpoons$ DMOS$^{+\cdot}$ + e
2 DMOS$^{+\cdot}$ $\xrightarrow{K2}$ (DMOS)$_2^{2+}$

$$(24)$$

It has been found that $K_2 > K_1$; hence, DMOS$^{+\cdot}$ dimerizes mainly by the radical route. In the presence of nucleophiles such as methanol and water, an irreversible reaction of DMOS$^{+\cdot}$ with nucleophiles (solvolysis) takes place together with the radical route. The relative contribution of the solvolysis route to the entire reaction depends on the concentration of the nucleophiles.

2.2.5 Oxidation of Enolic Olefins

Enolic olefins, i.e. olefins bearing electron-donating substituents such as alkoxy, acyloxy, and dialkylamino groups are oxidizable by the anodic method (Table 8) [1, 19].

Table 8. Oxidation Potentials of Some Enol Ethers and Acetates

Enol ether	V^a (vs. Ag/Ag$^+$)	Enol acetate	V^b (vs. SCE)
⟍OC$_2$H$_5$	1.72	⟍OAc	1.82
⬡O	1.48	⬡OAc	1.63
⬡—OC$_2$H$_5$	1.28, 1.63	⟍OAc	1.83
⬠—OC$_2$H$_5$	1.25, 1.64	AcO⟍(phenyl)	1.59
⟍OC$_2$H$_5$	1.27	⬡—OAc	1.93
(phenyl)⟍OCH$_3$	1.20	⬠—OAc	1.48

^{a} $6 \cdot 10^{-3}$ M, 0.5 M NaClO$_4$, CH$_3$OH
^{b} $2 \cdot 10^{-4}$ M, 0.1 M LiClO$_4$, CH$_3$CN

The addition of methoxy groups to the unsaturated bond easily takes place by anodic oxidation of enol ethers in methanol containing sodium methoxide; yields are generally satisfactory [20, 21].

$$CH_2{=}CHOR \xrightarrow[CH_3OH]{-2e} CH_3OCH_2CH(OR)OCH_3 \qquad (25)$$

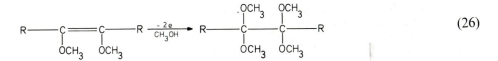

$$(26)$$

2. Anodic Oxidations

The anodic oxidation of enol ethers at a graphite anode in methanol containing 2,6-lutidine and sodium perchlorate results in the dimerization of the enol ethers to acetals of 1,4-dicarbonyl compounds [19].

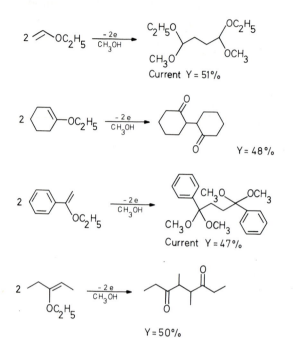

Current Y = 51%

Y = 48%

(27)

Current Y = 47%

Y = 50%

The mechanism of dimerization involves a tail-to-tail coupling of the cation radicals generated by the one-electron oxidation of enol ethers.

When the oxidative dimerization is applied to mixed solutions of enol ethers with styrene or α-methylstyrene in methanol containing sodium iodide or sodium perchlorate, mixed dimers are obtained together with symmetrical dimers and dimethoxylated monomers [22].

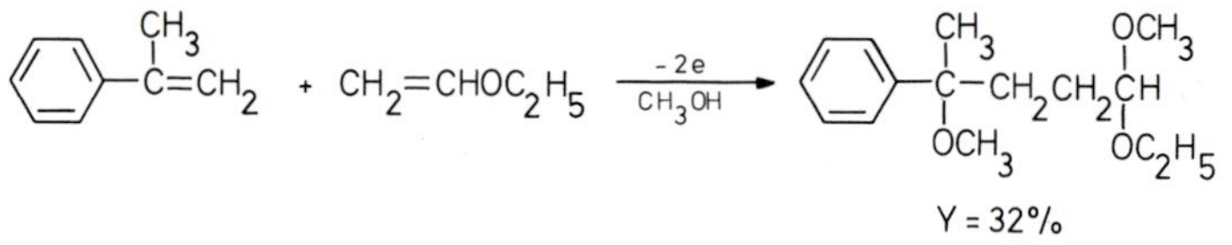

(28)

Y = 32%

Compared with enol ethers, enol acetates may more easily be prepared from the corresponding carbonyl compounds, and their oxidation potentials are generally in the region accessible by the anodic method. The oxidation of enol acetates in acetic acid containing tetraethylammonium *p*-toluene-sulfonate gives four types of compounds: conjugated enones (A), α-acetoxy-carbonyl compounds (B), *gem*-diacetoxy compounds (C), and triacetoxy compounds (D) [1].

22

Similar to enol ethers, the first reactive intermediates are cation radicals generated from enol acetates by one-electron oxidation (29). The yields and the distribution of products A, B, C, and D depend on the structure of the enol acetates and reaction conditions [23].

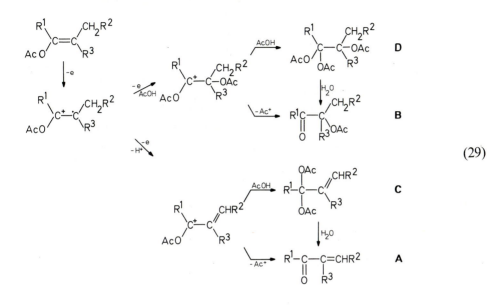

(29)

Table 9. Effect of Supporting Electrolytes

Enol acetate	Supporting electrolyte	Yield (%) Product A		Product B	
OAc, C_6H_{13}	$(C_2H_5)_4NOTs$ AcOK	C_5H_{11}	6 0	C_6H_{13}, OAc	24 78
OAc	$(C_2H_5)_4NOTs$ AcOK	=O	3 0	=O, OAc	37 82
OAc	$(C_2H_5)_4NOTs$ AcOK	=O	9 0	=O, OAc	25 88
OAc	$(C_2H_5)_4NOTs$ AcOK	O	90 25	O, OAc	0 60
OAc	$(C_2H_5)_4NOTs$ AcOK	=O	90 24	=O, OAc	0 59

23

2. Anodic Oxidations

The existence of the α-alkyl substituents R³ on the starting enol acetates is one of the main factors controlling the relative yields of two main products A and B. For instance, when using tetraethylammonium *p*-toluenesulfonate as the supporting electrolyte, acyclic and α-nonsubstituted alicyclic enol acetates are preferentially oxidized to α-acetoxyketones (B) whereas α-alkylated alicyclic enol acetates yield exclusively α,β-unsaturated enones (A).

Furthermore, the supporting electrolyte also exerts a significant influence on the relative rates of these competitive reaction pathways. As shown in Table 9, the use of potassium acetate (or triethylamine) instead of tetraethylammonium *p*-toluenesulfonate as the supporting electrolyte generally brings about a considerable increase in the yield of α-acetoxyketones (B).

This remarkable effect of the supporting electrolyte is attributable to the higher concentration of the acetate ions which may cause an increase in the rate of the solvolysis of the intermediate cationic species rather than in the rate of proton elimination.

The formation of α,β-unsaturated enones from enol acetates has been applied to the synthesis of 2,3-disubstituted 2-cyclopentenones including jasmone homologs (30).

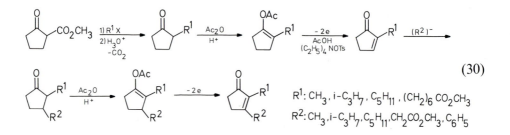

(30)

R¹: CH_3, i-C_3H_7, C_5H_{11}, $(CH_2)_6 CO_2CH_3$
R²: CH_3, i-C_3H_7, C_5H_{11}, $CH_2CO_2CH_3$, C_6H_5

The yields of the anodic oxidation are usually in the range of 80–90%.

This anodic oxidation has also been applied to the transformation of *l*-menthone to *d*-menthone (31) [24].

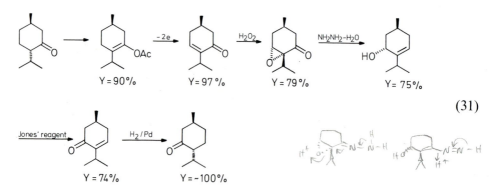

(31)

24

The effect of the solvent system on the product distribution in the anodic oxidation of enol acetates has been demonstrated by the synthesis of *l*-carvon from *l*-α-pinene [25].

The anodic oxidation of an enol acetate derived from α-pinene in acetic acid or methanol yields a mixture of products (32).

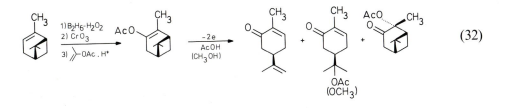

$$(32)$$

However, the use of a mixed solvent (8:1) of methylene chloride and acetic acid containing tetraethylammonium *p*-toluenesulfonate as the supporting electrolyte leads to the exclusive formation of pure *l*-carvone in 64% yield.

This anodic α-acetoxylation or α-methoxylation of ketones has been shown to be a powerful tool for the 1,2-transposition of the carbonyl group of ketones. The overall process is described by Eq. (33) [26].

$$(33)$$

Two typical examples of such a transposition are illustrated in scheme (34), though this transposition is really versatile.

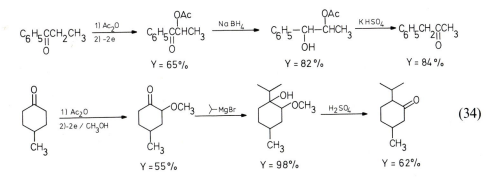

$$(34)$$

2. Anodic Oxidations

The concept of this 1,2-transposition can be extended to 1,4-transposition by using enones as the starting compounds. The anodic methoxylation of dienol acetates prepared from enones in a mixed solvent of acetic acid and methanol (1:9) yields γ-methoxylated enones regioselectively. Reduction of the γ-methoxylated enones with NaBH$_4$ to the corresponding alcohols followed by solvolysis of the tosylates of these alcohols in aqueous acetone gives products in which the carbonyl group is transposed to the γ-position of the starting enones [27].

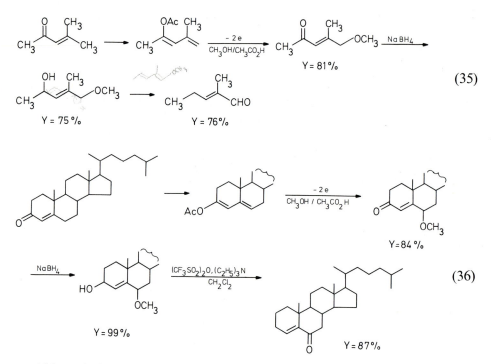

(35)

(36)

Although the oxidative dimerization of enol acetates is not known so far, the anodic oxidation of 1-acetoxy-1,6-heptadiene derivatives yields intramolecular cyclized products, cyclohexenyl ketones, as the main products, suggesting that the cyclization takes place through the electrophilic attack of the double bond by the cationic center generated from the enol ester moiety (37) [28].

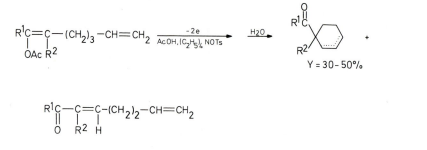

(37)

The anodic oxidation of enamines in methanol containing sodium methoxide as the supporting electrolyte shows a reaction pattern different from that of enol ethers or enol acetates. The main products are mixtures of isomeric methoxylated enamines (Y = 74–76%) (38) [20 b, 29].

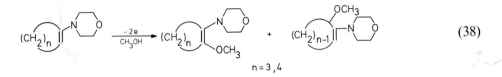

$$n = 3, 4 \tag{38}$$

Similarly, the oxidation of enamines in the presence of anions generated from methyl acetoacetate, acetylacetone, or dimethyl malonate also yields mixtures of substituted enamines (39) [30].

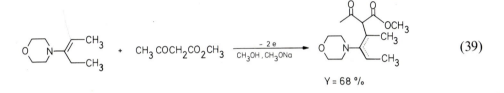

$$Y = 68\% \tag{39}$$

Enamines possessing an electron-withdrawing substituent yield pyrrole derivatives through oxidative dimerization (40) [31].

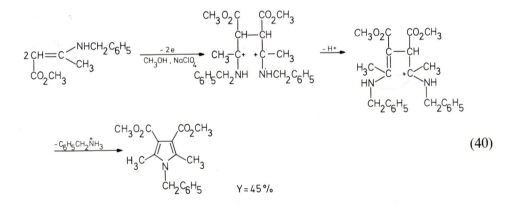

$$Y = 45\% \tag{40}$$

Aromatization of 1,4-dihydropyridine derivatives by anodic oxidation has been observed (41) [32].

$$\tag{41}$$

27

References

1. Shono, T., Matsumura, Y., Nakagawa, Y.: J. Am. Chem. Soc. *96*, 3532 (1974)
2. Shono, T., Ikeda, A.: J. Am. Chem. Soc. *94*, 7892 (1972)
3. a) Shono, T., Kosaka, T.: Tetrahedron Lett. *1968*, 6207
   b) Baggaley, A. J., Brettle, R.: J. Chem. Soc. (C) *1968*, 2055
   c) Fiaita, G., Fleischmann, M., Pletcher, D.: J. Electroanal. Chem. *25*, 455 (1970)
   d) Brettle, R., Sutton, J. R.: J. Chem. Soc., Perkin Trans. (I) *1975*, 1947
   e) Yoshida, K., Kanbe, T., Fueno, T.: J. Org. Chem. *42*, 2313 (1977)
4. Shono, T., Ikeda, \., Kimura, Y.: Tetrahedron Lett. *1971*, 3599
5. Bewick, A., Mellor, J. M., Pons, B. S.: J. Chem. Soc., Chem. Commun. *1978*, 738
6. Adams, C., Frankel, E. N., Utley, J. H. P.: J. Chem. Soc., Perkin Trans. I, *1979*, 353
7. Clennan, E. C., Simmons, W., Almgren, C. W.: J. Am. Chem. Soc. *103*, 1098 (1981)
8. Baltes, H., Steckhan, E., Schäfer, H. J.: Chem. Ber. *111*, 1294 (1978)
9. Shono, T., Ikeda, A.: Chem. Lett. *1976*, 311
10. Shono, T., Nishiguchi, I., Okawa, M.: Chem. Lett. *1976*, 573
11. Baltes, H., Stork, L., Schäfer, H. J.: Justus Liebigs Ann. Chem. *1979*, 318
12. a) Shono, T., Ikeda, A., Hayashi, J., Hakozaki, S.: J. Am. Chem. Soc. *97*, 4261 (1975)
    b) Shono, T., Ikeda, A., Hakozaki, S.: Tetrahedron Lett. *1972*, 4511
13. Zinger, B., Becker, J. Y.: Electrochimica Acta *25*, 791 (1980)
14. Eberson, L., Schäfer, H. J.: Organic Electrochemistry, Berlin—Heidelberg— New York, Springer 1971
15. Engels, R., Schäfer, H. J., Steckhan, E.: Liebigs Ann. Chem. *1977*, 204
16. See also Cedhein L., Eberson, L.: Acta Chem. Scand. *B30*, 527 (1976)
17. a) Steckhan, E., Schäfer, H. J.: Angew. Chem. *86*, 480 (1974)
    b) See also Sternerup, H.: Acta Chem. Scand. *B28*, 579 (1974)
18. a) Steckhan, E.: J. Am. Chem. Soc. *100*, 3526 (1978)
    b) Burgbacher, G., Schäfer, H. J.: J. Am. Chem. Soc. *101*, 7590 (1979)
    c) See also Eberson, L., Parker, V. D.: Acta Chem. Scand. *24*, 3563 (1970)
19. Koch, D., Schäfer, H. J., Steckhan, E.: Chem. Ber. *107*, 3640 (1974)
20. a) Belleau, B., Au-Young, Y. K.: Can. J. Chem. *47*, 2117 (1969)
    b) Shono, T., Matsumura, Y., Hamaguchi, H.: Bull. Chem. Soc. Jpn. *51*, 2179 (1978)
21. a) Shono, T., Toda, T., Oda, R.: Nippon Kagaku Zasshi *90*, 1260 (1969)
    b) Couture, R., Belleau, B.: Can. J. Chem. *50*, 3424 (1972)
    c) Michel, M. A., Martigny, P., Simonet, J.: Tetrahedron Lett. *1975*, 3143
22. Schäfer, H. J., Steckhan, E.: Tetrahedron Lett. *1970*, 3835
23. Shono, T., Okawa, M., Nishiguchi, I.: J. Am. Chem. Soc. *97*, 6144 (1975)
24. Shono, T., Matsumura, Y., Hibino, K., Miyawaki, S.: Tetrahedron Lett. *1974*, 1295
25. Shono, T., Nishiguchi, I., Yokoyama, T., Nitta, M.: Chem. Lett. *1975*, 433
26. Shono, T., Nishiguchi, I., Nitta, M.: Chem. Lett. *1976*, 1319
27. Shono, T., Kashimura, S.: J. Org. Chem., *48*, 1939 (1983)
28. Shono, T., Nishiguchi, I., Kashimura, S., Okawa, M.: Bull. Chem. Soc. Jpn. *51*, 2181 (1978)

29. See also a) Huang, S. J., Hso, E. T.: Tetrahedron Lett. *1971*, 1385
    b) Fritsch, J. M., Weingarten, H., Wilson, J. D.: J. Am. Chem. Soc. *92*, 4038 (1970)
30. Chiba, T., Okimoto, M., Nagai, H., Takato, Y.: J. Org. Chem. *44*, 3519 (1979)
31. Koch, von D., Schäfer, H. J.: Angew. Chem. *85*, 264 (1973)
32. a) Pragst, F., Kaltofen, B., Volke, J., Kuthan, J.: J. Electroanal. Chem. *119*, 301 (1981)
    b) See also Ludvik, J., Klima, J., Volke, J., Kurfürst, A., Kuthan, J.: J. Electroanal. Chem. *138*, 131 (1982)

2.3 Oxidation of Alcohols, Glycols, Ethers, and Acetals

2.3.1 Oxidation of Alcohols

The direct anodic oxidation of aliphatic saturated alcohols to the corresponding carbonyl compounds is not always effective, because the extremely high oxidation potentials of these alcohols make the direct removal of an electron from the lone-pair electrons on the oxygen atom difficult (Table 1) [1]. The indirect oxidation of alcohols is described in section 2.9.

The direct electrochemical oxidation of alcohols has been surveyed by D. C. Scholl *et al.* [2], and their conclusions are that this oxidation is best achieved in the neat liquid substrate, acetonitrile is the best solvent, a fluoroborate as the supporting electrolyte is recommended to obtain higher yields, and oxidation with a controlled potential is not effective.

Two patterns of reaction mechanism have been proposed for the oxidation of alcohols under neutral conditions. The first mechanism involves

Table 1. Oxidation Potentials of Alcohols[a]

Alcohol	$E_{1/2}\left(vs.\ \dfrac{ferrocene}{ferrocinium}\right)^b$
CH_3OH	2.73
C_2H_5OH	2.61
C_3H_7OH	2.56
$sec-C_3H_7OH$	2.50
C_4H_9OH	2.56
$sec-C_4H_9OH$	2.55
$C_5H_{11}OH$	2.46

[a] CH_3CN, supporting electrolyte 0.15 M $(C_4H_9)_4NBF_4$
[b] These values are almost the same as $E_{1/2}$ *vs.* Ag/0.01 M Ag^+

removal of one electron from the lone-pair electrons of the oxygen atom as the initiation step (1)-1), while the initiation process in the second mechanism involved abstraction of a hydrogen atom from the position α to oxygen (1)-2). Under strongly basic conditions, the oxidation of the alkoxide anion is the initiation step (1)-3).

$$1) \quad RCH_2OH \xrightarrow{-e} RCH_2\overset{+\cdot}{OH} \xrightarrow{-H^+} \underset{\text{radical I}}{RCH_2O\cdot}$$

$$(1)$$

$$2) \quad RCH_2OH \xrightarrow[Y\cdot]{-H\cdot} \underset{\text{radical II}}{R\overset{\cdot}{C}HOH} \xrightarrow[-H^+]{-e} RCHO$$

$$3) \quad RCH_2O^- \xrightarrow{-e} RCH_2O\cdot \qquad\qquad \text{Y: radical species generated in solution or on the anode surface by the anodic oxidation.}$$

The fact that acetonitrile is the best solvent and fluoroborate the most suitable supporting electrolyte may suggest the importance of the first mechanism. The reactions occurring after the formation of the radical I are however not simple, because the radical I may behave like radical Y$\cdot$ as shown by the second mechanism. It can also abstract a hydrogen atom from another radical I to yield the corresponding aldehyde (2).

$$2\,RCH_2O\cdot \longrightarrow RCH_2OH \;+\; RCHO \qquad\qquad (2)$$

Thus, it may be assumed that the anodic oxidation of alcohols involves mechanism (1)-1) and some radical mechanism including mechanism (1)-2), and that the relative contribution of each mechanism to the entire reaction depends on the structure of the alcohol and the reaction conditions. The formation of acetals has often been observed in the anodic oxidation of alcohols, though it has been explained by a non-electrochemical mechanism. One example of the oxidation of alcohols is shown below.

$$C_4H_9OH \xrightarrow[LiBF_4]{-2e} C_3H_7CHO \qquad\qquad (3)$$

neat   platinum anode   Y = 77%

The assumption that the surface of metallic anode is involved in the oxidation of alcohols as a hydrogen abstractor has been confirmed by the oxidation of primary alcohols to the corresponding carboxylic acids using nickel hydroxide as the anode [3]. The anode is prepared before each electrolysis by treatment of nickel with a low-frequency alternating current in 0.1 N $NiSO_4$ — 0.1 N CH_3CO_2Na — 0.005 N NaOH [4]. The oxidation is carried out in t-C_4H_9OH—H_2O—KOH (B) or in aqueous NaOH (A) [5]. Short-chain alcohols are oxidized at room temperature whereas higher temperatures are necessary for the oxidation of longer-chain alcohols. The oxidation process involves abstraction of a hydrogen from the α-carbon of alcohols by nickel peroxide [$Ni^{iii}O(OH)$] formed on the anode surface by the oxidation (4), though this detailed mechanism of the formation of aldehydes and carboxylic acids is not necessarily clear.

$$R-CH_2OH \xrightarrow{Ni^{iii}} R-\dot{C}HOH \xrightarrow[-H^+]{-e} R-CHO \longrightarrow R-CO_2H \tag{4}$$

Some typical results are listed in Table 2.

Table 2. Oxidation of Primary Alcohols to Carboxylic Acids on a Nickel Hydroxide Anode

R in RCH$_2$OH	Reaction system	Yield (%)	R in RCH$_2$OH	Reaction system	Yield (%)
C$_3$H$_7$	B	92	CH$_3$O—⟨⟩—CH$_2$	B	90[a]
C$_5$H$_{11}$	B	91			
C$_6$H$_{13}$	A	84	Cl—⟨⟩—CH$_2$	B	71[a]
C$_7$H$_{15}$	A	89	⟨⟩—CH$_2$CH$_2$	B	70[a]
C$_8$H$_{17}$	A	89			
C$_9$H$_{19}$	A	87	CH$_3$O—⟨⟩—	B	83[b]
C$_{11}$H$_{23}$	A	80	HO—⟨⟩—	B	67[b]
C$_{17}$H$_{35}$	A	77			
⟍⟍CH	A	73	⟨O⟩	A	79
⟍⟍⟍CH	A	76	⟩=CH	A	10
CH≡C	A	51	⟍⟍CH$_2$	B	34
⟨⟩—	A	86	⟍⟍⟍CH$_2$	A	82
O$_2$N—⟨⟩—	B	91[a]	⟍⟍⟍⟍CH$_2$	A	68

[a] See ref. [3d]; [b] Aldehydes

2.3.2 Oxidation of Glycols

Although the anodic oxidation method is not always suitable for the oxidation of simple aliphatic alcohols, it is highly efficient in the oxidative cleavage of glycols and related compounds. The anodic oxidation of glycols and glycol ethers in methanol containing tetraethylammonium p-toluenesulfonate as a supporting electrolyte results in a clean cleavage of glycols to the corresponding carbonyl compounds (5) [6].

$$R^1-\underset{\underset{OR}{|}}{\overset{\overset{R^2}{|}}{C}}-\underset{\underset{OR}{|}}{\overset{\overset{R^3}{|}}{C}}-R^4 \xrightarrow[CH_3OH]{-2e} R^1R^2C=O \;+\; R^3R^4C=O \qquad\qquad (5)$$
$$R=H, CH_3$$

This anodic oxidation does not show any of the stereochemical limitations usually observed in cleavage reaction by chemical oxidizing reagents [7]. Furthermore, 1,2-dimethoxy- and 1-hydroxy-2-methoxy-alkanes are also oxidized with almost similar current efficiencies.

Cyclohexene oxide may also be oxidized, the initial step involving formation of the corresponding hydroxy ether.

The initiation step of this anodic oxidation of glycols may be the electron transfer from the lone-pair electrons of the oxygen atom to the anode.

This transfer may be assisted by the neighboring hydroxy or alkoxy group, since 1,3- or 1,4-dihydroxycyclohexane, 2-methylcyclohexanol or 3,3-dimethylbutan-2-ol are not anodically oxidized under the same reaction conditions. A similar electrochemical cleavage has also been observed in the oxidation of a vicinal diacetal [10].

The anodic cleavage of 1,2-glycols has been utilized for a variety of organic syntheses, e.g. for the synthesis of symmetrical ketones from methyl methoxyacetate (7) [11].

$$CH_3OCH_2CO_2CH_3 \xrightarrow{2\,RMgX} CH_3OCH_2\underset{\underset{R}{|}}{\overset{\overset{R}{|}}{C}}-OH \xrightarrow{-2e} R_2C=O \qquad\qquad (7)$$
$$Y = 80-88\%$$
$$R=C_3H_7,\ iso-C_3H_7,\ C_4H_9,\ iso-C_4H_9,\ cyclo-C_6H_{11}$$

Unsymmetrical ketones are also synthesized from symmetrical ketones, the anodic cleavage of 1,2-amino alcohols [12] representing the key reaction (8) [11].

$$RCH_2-\overset{\overset{O}{\|}}{C}-CH_2R \longrightarrow RCH_2-\overset{\overset{O}{\|}}{C}-\underset{\underset{\underset{H_5C_2}{\diagup}\,\underset{C_2H_5}{\diagdown}}{N}}{C}HR \xrightarrow{R'MgX} RCH_2-\underset{\underset{\underset{R'}{|}}{\overset{\overset{OH}{|}}{C}}-\underset{\underset{\underset{H_5C_2}{\diagup}\,\underset{C_2H_5}{\diagdown}}{N}}{C}HR \xrightarrow{-2e} RCH_2-\overset{\overset{O}{\|}}{C}-R'$$

$$Y = 52-72\%$$
$$R=H, C_2H_5;\quad R'= iso-C_3H_7, C_4H_9,\ cyclo-C_6H_{11}$$

$$(8)$$

Table 3. Anodic Oxidation of 1,2-Diols and Related Compounds

Starting compound	Product and distribution (%)		Total yield (%)
cyclopentane-1,2-diol (–OH, –OH)	$(CH_2)_3$ $\big<$ $CH(OCH_3)_2$ / $CH(OCH_3)_2$ (A, 56)	$(CH_2)_3$ $\big<$ $CH(OCH_3)_2$ / CHO (B, 14)	96
	CH_3O–(ring O)–OCH_3 (C, 26)		
cyclopentane-1,2-diol	(A, 56), (B, 20), (C, 18)		94
cyclopentane-1,2-di(OCH$_3$)	(A, 65), (B, 18)		67
cyclohexane-1,2-diol (–OH, –OH)	$(CH_2)_4$ $\big<$ $CH(OCH_3)_2$ / $CH(OCH_3)_2$ (D, 51)	$(CH_2)_4$ $\big<$ $CH(OCH_3)_2$ / CHO (E, 14)	65
cyclohexane epoxide	(D, 43), (E, 9)		52
1-methylcyclohexane-1,2-diol	open-chain diketone (36)	CH_3O OCH_3 substituted ketone (9)	61
	cyclopentene with $C(=O)CH_3$ and CH_3 (16)		
HO– cyclopentane –CO$_2$CH$_3$ / HO– –CO$_2$CH$_3$	CH_3O, CH_3O / CH_3O, OCH_3 with CO_2CH_3, CO_2CH_3		83[a]
$(CH_3)_2C(OH)$–$C(OH)(CH_3)_2$	$(CH_3)_2C(OCH_3)_2$ (11)	$(CH_3)_2C{=}O$ (78)	89
tetrahydrofuran-2-yl-CH(OH)CH$_3$	furan ring with OCH_3		76[b]

[a] Reaction system: $CH_3OH{-}H_2SO_4$ [8]
[b] See ref. [9]

As described in Section 2.2.5, the anodic oxidation of enol ethers in methanol yields α-methoxylated carbonyl compounds, which are useful starting compounds in the synthesis of carbonyl compounds utilizing the technique of oxidative cleavage of glycols.

2. Anodic Oxidations

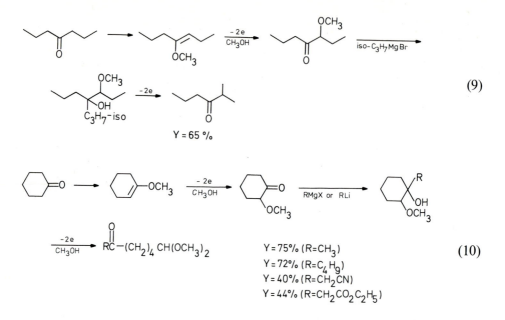

(9)

(10)

The first products in the anodic oxidation of enol ethers in methanol have been known to be dimethyl acetals of α-methoxylated carbonyl compounds which, upon further oxidation, yield the oxidative cleavage products [13].

(11)

Enol acetates and α-hydroxy ketones have also been cleaved by anodic oxidation [14].

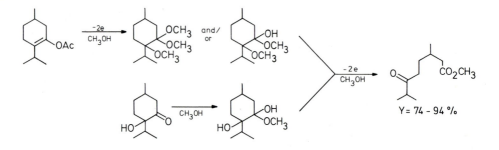

(12)

34

In contrast to the facile oxidation of glycol ethers, saturated aliphatic ethers are oxidized rather reluctantly. The anodic oxidation of the latter in methanol containing sodium methoxide, tetraethylammonium p-toluene-sulfonate, or ammonium nitrate as the supporting electrolyte yields the corresponding α-methoxylated ethers, though yields are not satisfactory so far [15].

Table 4. Anodic Methoxylation of Ethers Using CH_3ONa as a Supporting Electrolyte

Ether	Product	Yield (%)
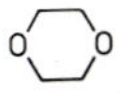 (1,3-dioxane)	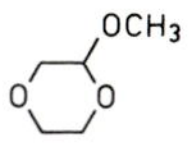 (2-methoxy-1,3-dioxane, OCH₃)	28
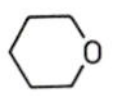 (tetrahydropyran)	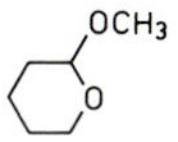 (2-methoxytetrahydropyran, OCH₃)	26
(tetrahydrofuran)	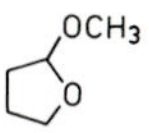 (2-methoxytetrahydrofuran, OCH₃)	16.3
(thiolane)	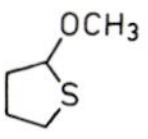 (2-methoxythiolane, OCH₃)	8.5
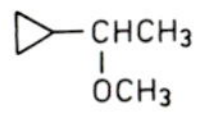 $\triangleright\!-\!CHCH_3$, OCH_3	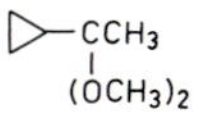 $\triangleright\!-\!CCH_3$, $(OCH_3)_2$	24.3
	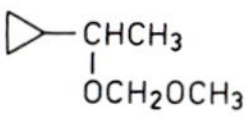 $\triangleright\!-\!CHCH_3$, OCH_2OCH_3	7.8
$C_2H_5{>}CH\!-\!OCH_3$, CH_3	$C_2H_5{>}C(OCH_3)_2$, CH_3	21

Similar to the oxidation of alcohols, two mechanisms are conceivable for the initiation step. One involves the direct removal of an electron from the lone-pair electrons of the oxygen atom (13) and the other abstraction of a hydrogen atom from the α-position of the ether by a radical species (S·) generated by the anodic oxidation of solvents or supporting electrolytes (14).

$$RCH_2OR' \xrightarrow{-e} RCH_2\overset{+\cdot}{O}R' \xrightarrow[-H^+]{-e} R\overset{+}{C}HOR' \tag{13}$$

$$RCH_2OR' \xrightarrow{S\cdot} R\dot{C}HOR' \xrightarrow{-e} R\overset{+}{C}HOR' \tag{14}$$

35

In general, no conclusive evidence of either of these reaction mechanism has been obtained so far.

The radical mechanism (14) seems to be favored on the basis of the following observations. a) No oxidation peak attributable to the direct oxidation of ethers has been observed. b) The controlled potential oxidation of dioxane at 1.65 V *vs.* SCE in methanol using ammonium nitrate as a supporting electrolyte yields the methoxylated product in a reasonable yield while no methoxylation occurs when ammonium nitrate is replaced by lithium perchlorate. Lithium perchlorate is stable at this potential whereas ammonium nitrate is oxidized to a radical species. c) Sodium methoxide, which seems to generate methoxy radical under these reaction conditions, always gives the highest yield.

On the other hand, the mechanism of the direct oxidation (13) is favored by the results of a study on isotope effects. Both intra- and intermolecular isotope effects in the anodic methoxylation of ethers have been determined [16] and compared with those obtained in the Kharasch-Sosnovsky reaction [17] and in the anodic oxidation of carbamates (Table 5) [18].

If the initiation step involves hydrogen abstraction by a radical species, both intra- and intermolecular isotope effects must be almost identical as shown for the Kharasch-Sosnovsky reaction in which the active radical species is a tertiary butoxy radical. On the other hand, in the direct oxidation, the intra- and intermolecular isotope effects would be different, and both effects are considered to be lower than those observed in radical processes like the Kharasch-Sosnovsky reaction since the intramolecular isotope effect is caused by the relative rate of the removal of a proton or deuterium cation from the highly exited cation radical species of ethers. The inter-

Table 5. Isotope Effects (K_H/K_D)

Compound	Anodic methoxylation (supporting electrolyte)	Kharasch-Sosnovsky reaction
	2.1 ($NaOCH_3$) 2.0 (($C_2H_5)_4NOTs$)	3.2
	1.5-1.6 ($NaOCH_3$) 1.6-1.7 (($C_2H_5)_4NOTs$)	3.1-3.2
	1.86 ± 0.05 (($C_2H_5)_4NOTs$)	
	1.59 ± 0.05 (($C_2H_5)_4NOTs$)	

36

molecular isotope effect may then be attributed to the slight difference in the oxidation potentials between ethers and deuterated ethers.

As shown in Table 5, both intra- and intermolecular isotope effects observed in the anodic methoxylation are reasonably lower than those occurring in the Kharasch-Sosnovsky reaction. Moreover, the intra- and intermolecular isotope effects are very similar to those obtained in the anodic methoxylation of carbamates which has been shown to proceed by the direct oxidation mechanism. Although it is not conclusive, the study on the isotope effects supports the mechanism of the direct oxidation which has also been suggested for the anodic oxidation of 2-methoxyethanol [19].

The electrochemical methoxylation of acetals has been shown to occur under conditions similar to those of the methoxylation of ethers (15) [20]. The highest yield is obtained when potassium hydroxide is used as the supporting electrolyte. Dioxolanes always give better results than dimethoxy-acetals.

$$
C_3H_7CH \overset{OCH_2}{\underset{OCH_2}{\Big|}} \quad \xrightarrow[CH_3OH]{-2e} \quad \underset{CH_3O}{\overset{C_3H_7}{\diagdown}} C \overset{OCH_2}{\underset{OCH_2}{\diagup}} \Big| \tag{15}
$$

$$Y = 72\%$$

References

1. a) Sundholm, G.: J. Electroanal. Chem. *31*, 265 (1971)
   b) Sundholm, G.: Acta Chem. Scand. *25*, 3188 (1971)
2. Scholl, P. C., Lentsch, S. E., Van De Mark, M. R.: Tetrahedron *32*, 303 (1976)
3. a) Vertes, G., Horanyi, G., Nagy, F.: Magy, Chem. Foly. *74*, 172 (1968); C.A. *68*, 110781 (1968)
   b) Vertes, G., Horanyi, G., Nagy, F.: Tetrahedron *28*, 7 (1972)
   c) Vertes, G., Horanyi, G., Nagy, F.: Croat. Chim. Acta *44*, 21 (1972)
   d) Amjad, M., Pletcher, D., Smith, C.: J. Electrochem. Soc. *124*, 203 (1977)
   e) Fleischmann, M., Korinek, K], Pletcher, D.: J. Electroanal. Chem. *31*, 39 (1971)
   f) Fleischmann, M., Korinek, K., Pletcher, D.: J. Chem. Soc. Perkin Trans. (2) *1972*, 1396
4. Briggs, G. W. D., Jones, E., Wynne-Jones, W. S. K.: Faraday Soc. *51*, 1433 (1955)
5. a) Kauleu, J., Schäfer, H. J.: Synthesis *1979*, 513
   b) See also Robertson, P. M.: J. Electroanal. Chem. *111*, 97 (1980)
6. Shono, T., Matsumura, Y., Hashimoto, T., Hibino, K., Hamaguchi, H., Aoki, T.: J. Am. Chem. Soc. *97*, 2546 (1975)
7. Angyal, S. J., Young, R. J.: J. Am. Chem. Soc. *81*, 5467 (1959)
8. Torii, S., Uneyama, K., Tanaka, H., Yamanaka, T., Yasuda, T., Ono, M.: J. Org. Chem. *46*, 3312 (1981)
9. Shono, T., Matsumara, Y., Hashimoto, T.: 29th Ann. Meeting Chem. Soc. Japan (1973), Abstract p. 953, Vol. II

10. a) Shono, T., Toda, T., Oda, R.: Nippon Kagaku Zasshi *90*, 1260 (1969)
    b) The same reaction is also reported in: Couture, R., Belleau, B.: Can. J. Chem. *50*, 3424 (1972)
    See also c) Kemula, W., Grakowski, Z. R., Kalinowski, M. K.: J. Am. Chem. Soc. *91*, 6863 (1969)
    d) Michielli, R. F., Elving, P. J.: J. Am. Chem. Soc. *91*, 6864 (1969)
11. Shono, T., Hamaguchi, H., Matsumura, Y., Yoshida, K.: Tetrahedron Lett. *1977*, 3625
12. See also Masui, M., Kamada, Y., Ozaki, S.: Chem. Pharm. Bull. *28*, 1619 (1980)
13. Shono, T., Matsumura, Y., Imanishi, T., Yoshida, K.: Bull. Chem. Soc. Jpn. *51*, 2179 (1978)
14. Torii, S., Inokuchi, T., Oi, R.: J. Org. Chem. *47*, 47 (1982)
15. Shono, T., Matsumura, Y.: J. Am. Chem. Soc. *91*, 2803 (1969)
16. Shono, T., Matsumura, Y.: unpublished
17. Rawlinson, D. J., Sosnovsky, G.: Synthesis *1972*, 1
18. Shono, T., Hamaguchi, H., Matsumura, Y.: J. Am. Chem. Soc. *97*, 4764 (1975)
19. Ross, S. D., Barry, J. E., Finkelstein, M., Rodd, E. J.: J. Am. Chem. Soc. *95*, 2193 (1973)
20. Scheeren, J. W., Goossens, H. J. M., Top, A. W. H.: Synthesis *1978*, 283

2.4 Oxidation of Compounds Containing Sulfur, Phosphorous, and Boron

2.4.1 Sulfur Compounds

Thiols are easily converted into disulfides by electrooxidation. The oxidation peak potential of thiophenol on a platinum electrode in aqueous methanolic solution (50% v/v) depends on pH. The plot of E_p against pH gives a slope of 60 mV/pH at pH 8.2. In alkaline medium, the oxidation peak potential is about 0.3 V *vs*. SCE. In DMF solution, two oxidation peaks are observed when a platinum anode is used. The peak at about 0 V *vs*. SCE corresponds to the oxidation of $C_6H_5S^-$ and the peak at about 1.1 V *vs*. SCE is attributable to the following reaction [1]:

$$2\ C_6H_5SH \xrightarrow[-2H^+]{-2e} C_6H_5SSC_6H_5 \tag{1}$$

Further oxidation of disulfides may be achieved on a platinum electrode in acetonitrile using sodium perchlorate as the supporting electrolyte [2]. Diphenyl disulfide shows two oxidation peaks at about 1.2 V *vs*. Ag/Ag$^+$ and about 1.5 V *vs*. Ag/Ag$^+$. At the first oxidation potential, one electron is removed from the sulfur atom to yield a cation radical (2) whereas details on the formation of the second peak are not known. The cation radical is not stable but changes to a cationic species which can be trapped by alkenes such as cyclohexene and 1-octene (3) [3].

$$C_6H_5SSC_6H_5 \xrightarrow{-e} (C_6H_5SSC_6H_5)^{+\cdot} \qquad (2)$$

$$(RSSR)^{+\cdot} \xrightarrow[CH_3CN]{} RS-\overset{+}{N}=\overset{\cdot}{C}CH_3 \xrightarrow[R'-CH=CH_2]{} R'\overset{\overset{\displaystyle NHCOCH_3}{|}}{C}H-CH_2SR \qquad (3)$$
$$R=C_6H_5, \quad R'=C_6H_{13} \quad Y=84\%$$

The oxidation of organic sulfides to sulfoxides or sulfones is easily achieved in high yields in aqueous solutions. Single sweep voltammetry of diphenyl sulfide with 0.18 M perchloric or sulfuric acid as the supporting electrolyte shows a sharp oxidation peak at 1.30 V vs. SCE [4]. The controlled potential oxidation of sulfides in perchloric acid at 1.10 V yields sulfoxide in almost quantitative current yield without any contamination with diphenyl sulfone. Dimethyl sulfide is completely oxidized to dimethyl sulfone in acetonitrile containing only 1 % of water [5]. In anhydrous acetonitrile containing sodium perchlorate as the supporting electrolyte, the final products are sodium methanesulfonate and carbon monooxide. The first step of the reaction involves one-electron transfer from the sulfide to form a cation radical which rapidly loses a proton and a second electron to yield a sulfonium derivative which, upon reaction with the starting sulfide, forms the dimethyl methyl-thiomethylsulfonium ion as the major immediate product (4).

$$CH_3SCH_3 \xrightarrow{-e} (CH_3SCH_3)^{+\cdot} \xrightarrow[-H^+]{-e} \left\{ \begin{array}{c} CH_3SCH_2{}^+ \\ \updownarrow \\ CH_3\overset{+}{S}=CH_2 \end{array} \right\} \xrightarrow{(CH_3)_2S} CH_3SCH_2\overset{+}{S}\overset{\displaystyle CH_3}{\underset{\displaystyle CH_3}{<}} \qquad (4)$$

In the anodic oxidation of 2,6-di-*endo*-norbonyl derivatives and the corresponding *exo,endo*-norbonyl derivatives it has been observed that certain groups near a sulfur atom can assist oxidation at sulfur (Table 1) [6]. Table 1 shows that the peak potentials of the studied *exo* derivatives and *endo* acids and esters are in the expected range of the oxidation potentials of alkyl sulfides. However, the peak potentials of the *endo* acid anion and *endo* alcohol are dramatically shifted toward more cathodic values suggesting that the carboxylate anion and hydroxymethyl group assist in the removal of electrons from the sulfur atom.

The formation of sulfonium ions in the oxidation of sulfides has been clearly demonstrated in the anodic oxidation of diphenyl sulfide in aceto-nitrile on a platinum electrode using sodium perchlorate as the supporting electrolyte [7]. Diphenyl sulfide shows three oxidation peaks at about 1.1 V (number of electrons, n = 0.97), 1.3 V (n = 1.50) and 1.6 V vs. Ag/Ag$^+$

2. Anodic Oxidations

Table 1. Anodic Oxidation of Norbornyl Derivatives

X	![structure 1] CH$_3$S X $E_p^{a)}$	![structure 2] CH$_3$S X $E_p^{a)}$
CO$_2$H	1.20	1.28
CO$_2$CH$_3$	1.21	1.29
CH$_2$OH	0.56	1.20
NHCO$_2$C$_2$H$_5$	0.98	1.20
CO$_2^-$	0.65	1.28

$^{a)}$Peak potential of the first oxidation in acetonitrile;
0.1M (C$_4$H$_9$)$_4$NClO$_4$, vs. Ag/0.1M Ag$^+$

(n = 1.98). The first and second oxidation steps have been proposed to proceed by Eqs. (5) and (6).

$$2\,C_6H_5SC_6H_5 \xrightarrow[-H^+]{-2e} (C_6H_5)_2\overset{+}{S}C_6H_4SC_6H_5 \qquad (5)$$

$$2\,(C_6H_5)_2\overset{+}{S}C_6H_4SC_6H_5 \xrightarrow[-H^+]{-2e} (C_6H_5)_2\overset{+}{S}C_6H_4\underset{\underset{C_6H_5}{|}}{\overset{+}{S}}C_6H_4SC_6H_4\overset{+}{S}(C_6H_5)_2 \qquad (6)$$

Three different mechanisms (schemes (7) to (9)) have been suggested for the formation of the first sulfonium ion.

$$2\,C_6H_5SC_6H_5 \xrightarrow{-2e} 2\,C_6H_5\overset{+\cdot}{S}C_6H_5$$

$$C_6H_5\overset{+\cdot}{S}C_6H_5 \rightleftharpoons C_6H_5S-\!\!\!\bigcirc\!\!\!\cdot \;+\; H^+ \qquad (7)$$

$$C_6H_5\overset{+\cdot}{S}C_6H_5 \;+\; C_6H_5S-\!\!\!\bigcirc\!\!\!\cdot \longrightarrow C_6H_5SC_6H_4\overset{+}{S}(C_6H_5)_2$$

$$C_6H_5SC_6H_5 \xrightarrow{-e} C_6H_5\overset{+\cdot}{S}C_6H_5 \xrightarrow{-e} C_6H_5S-\!\!\!\bigcirc\!\!\!^+ \;+\; H^+ \qquad (8)$$

$$C_6H_5SC_6H_5 \;+\; C_6H_5S-\!\!\!\bigcirc\!\!\!^+ \longrightarrow C_6H_5SC_6H_4\overset{+}{S}(C_6H_5)_2$$

$$C_6H_5SC_6H_5 \xrightarrow{-e} C_6H_5\overset{+\cdot}{S}C_6H_5$$

$$C_6H_5\overset{+\cdot}{S}C_6H_5 \;+\; C_6H_5SC_6H_5 \longrightarrow C_6H_5\overset{+}{S}\!\!\left\langle \bigcirc \right\rangle\!\!-SC_6H_5 \xrightarrow{-e} C_6H_5SC_6H_4\overset{+}{S}(C_6H_5)_2 \;+H^+$$
$$\underset{\displaystyle C_6H_5}{|}$$

$$(9)$$

A different reaction pathway involving disproportionation of the cation radical has been proposed for the formation of the trianisylsulfonium cation by the anodic oxidation of dianisyl sulfide in the presence of anisole (10) [8].

$$(CH_3OC_6H_4)_2S \xrightarrow{-e} (CH_3OC_6H_4)_2\overset{+\cdot}{S}$$

$$2(CH_3OC_6H_4)_2\overset{+\cdot}{S} \longrightarrow (CH_3OC_6H_4)_2S \;+\; (CH_3OC_6H_4)_2S^{2+} \qquad\qquad (10)$$

$$(CH_3OC_6H_4)_2S^{2+} \;+\; CH_3OC_6H_5 \longrightarrow (CH_3OC_6H_4)_3\overset{+}{S} \;+\; H^+$$

The cation radical or some other cationic species formed from diphenyl sulfide interact with alcohols. Thus, the anodic oxidation of a solution of diphenyl sulfide and alcohol in methylene dichloride gives the corresponding alkyl chloride in reasonable yield [9]. Although the mechanism of this reaction is not always clear, the following mechanism seems the most reasonable one; the cation radical of diphenyl sulfide reacts with the alcohol to form sulfonium ion which yields the corresponding alkyl chloride through reaction with the chloride anion formed *in situ* from methylene chloride (11).

$$(C_6H_5)_2S \xrightarrow{-e} (C_6H_5)_2\overset{+\cdot}{S} \xrightarrow[\;2)\,-e\,,\,-H^+\;]{1)\;ROH} (C_6H_5)_2\overset{+}{S}-OR \xrightarrow{Cl^-} (C_6H_5)_2SO \;+\; RCl$$
$$Y=95\%\;\;(R=C_8H_{17})$$
$$Y=90\%\;\;(R=C_6H_5CH_2CH_2CH_2)$$

$$(11)$$

The S_N2-like character of the reaction of chloride anion with sulfonium salts has been clearly demonstrated by the almost complete inversion of configuration observed in the transformation of S-(+)-2-octanol to R-(−)-2-chlorooctane (12).

$$\xrightarrow[(C_6H_5)_2S\,,\,CH_2Cl_2]{-2\,e}$$

$$Y=75\%$$

(S(+), optical purity 98.5%)   (R(−), optical purity 94.6%)

$$(12)$$

2. Anodic Oxidations

Some 1,4- and 1,5-diols are converted into the corresponding cyclic ethers under the same reaction conditions (13).

$$(13)$$

The anodic oxidation of alcohols to the corresponding carbonyl compounds using sulfides is described in the Section 2.9.3.

Anodic cleavage of a carbon-sulfur bond has been observed in the oxidation of bis(phenylthio)methane in acetonitrile at a platinum electrode using sodium or tetraethylammonium perchlorate as the supporting electrolyte. This compound shows two oxidation peak potentials at 1.46 and 1.57 V *vs.* SCE at low sweep rates and is converted into diphenyl disulfide and formaldehyde through the controlled potential oxidation at 1.38 V (14) [10].

$$
\begin{array}{c}
C_6H_5S \\
C_6H_5S
\end{array}\!\!>\!\!CH_2 \xrightarrow[H_2O,\,-2H^+]{-2\,e} C_6H_5SSC_6H_5 + CH_2O \qquad (14)
$$

The anodic cleavage of alicyclic thioacetals has been shown to follow an EEC-type mechanism (15) when the reaction is carried out in non-nucleophilic media.

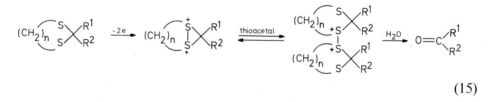

$$(15)$$

In the presence of stronger nucleophiles in large excess, however, the reaction follows an ECE-type mechanism (16) which is more suitable to explain the anodic degradation of 1,3-dithianes [11].

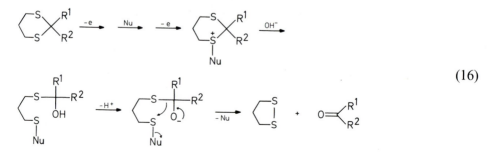

$$(16)$$

The fact that diphenyl sulfoxide, the primary oxidation product of diphenyl sulfide in acetonitrile, shows an anodic peak at 1.83 V $vs.$ Ag/Ag$^+$ where one electron per sulfoxide molecule is involved in this oxidation and that the yield of diphenyl sulfone is always about 50% suggests that the main oxygen source required for the oxidation must be diphenyl sulfoxide itself according to Scheme (17) [12].

$$C_6H_5SOC_6H_5 \xrightarrow{-e} C_6H_5\overset{+\cdot}{S}OC_6H_5$$

$$C_6H_5SOC_6H_5 + C_6H_5\overset{+\cdot}{S}OC_6H_5 \longrightarrow \underset{C_6H_5}{\overset{C_6H_5}{S}}\!\!>\!\!\overset{+}{S}\!-\!O\!-\!\underset{C_6H_5}{\overset{C_6H_5}{\underset{|}{\overset{|}{S}}}}\!-\!O\cdot \longrightarrow \begin{matrix} C_6H_5\overset{+\cdot}{S}C_6H_5 \\ + \\ C_6H_5SO_2C_6H_5 \end{matrix} \tag{17}$$

Each sulfoxide cation radical, the primary electron transfer product, reacts with one sulfoxide molecule giving one molecule of sulfone and a new sulfide cation radical. This reaction mechanism is in agreement with the result that only half of the sulfoxide is oxidized to sulfone. In the benzene-acetonitrile medium, the cation radical $C_6H_5\overset{+\cdot}{S}C_6H_5$ is trapped by benzene to yield triphenylsulfonium ion according to Eq. (18).

$$C_6H_5\overset{+\cdot}{S}C_6H_5 + C_6H_6 \xrightarrow{-e} (C_6H_5)_3 S^+ + H^+ \tag{18}$$

In pure acetonitrile, the cation radical reacts with acetonitrile giving another cation radical intermediate which, in turn, undergoes further anodic oxidation. Thus the overall reaction mechanism is consistent with the coulometric results.

The electrochemical oxidation of tetraalkylthioureas has been shown to proceed according to Eq. (19) in which the first step is the formation of a cation radical [13].

$$\overset{}{\underset{}{>}}C\!=\!S \xrightarrow{-e} [\!>\!C\!=\!S]^{+\cdot} \begin{cases} \xrightarrow{>C=S} \;\; \overset{+}{>}C\!-\!S\!-\!S\!-\!C\!<\!\!< \\ \qquad\qquad\downarrow{-e} \\ \qquad\;\; \overset{+}{>}C\!-\!S\!-\!S\!-\!C\!\!<_{+} \end{cases} \tag{19}$$

Dimerization of the cation radical takes place either directly by coupling of two cation radicals or addition of the cation radical to a neutral thiocarbonyl molecule followed by one-electron oxidation of the resultant dimeric cation radical. The cyclic voltammetry reveals that the latter process is the dominant pathway for the production of the dicationic species.

2.4.2 Compounds Containing Phosphorous and Boron

Phosphorous-containing compounds such as trialkyl- and triarylphosphines and trialkyl phosphites are rather easily oxidized to their cation radicals by the anodic method. Since a variety of reactions have already been reported, only one typical reaction is shown below [14].

$$(RO)_3P \xrightarrow{-e} (RO)_3\overset{+\cdot}{P} \xrightarrow{ArH} [(RO)_3P-ArH]^{+\cdot} \xrightarrow{-e,-H^+} (RO)_3\overset{+}{P}-Ar \xrightarrow{NaI} (RO)_2\underset{\underset{O}{\|}}{P}Ar$$

$$(20)$$

In contrast to boron tetrafluoride anion, tetraphenylboride anion $[B(C_6H_5)_4^-]$ is oxidized at a platinum electrode in acetonitrile at a potential of 0.92 V (peak potential *vs.* SCE), and tetrabutylboride anion $[B(C_4H_9)_4^-]$ is oxidized at 0.35 V. The mechanisms of the anodic oxidation of both anions are however different [15]. In the oxidation of the latter anion, the formation of free butyl radical has been observed by ESR technique whereas free phenyl radical has not been detected in the oxidation of the former anion though biphenyl has been isolated as an oxidation product. An intramolecular mechanism has been proposed for the formation of biphenyl by analyzing the products obtained by the anodic oxidation of a mixture of $(CH_3)_4N^+B^-(C_6H_5)_4$ and $(CH_3)_4N^+B^-(C_6D_5)_4$ (21) [16].

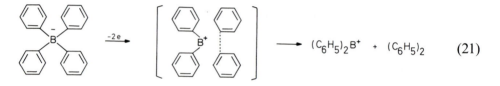

$$\xrightarrow{-2e} \qquad \longrightarrow (C_6H_5)_2B^+ \; + \; (C_6H_5)_2 \qquad (21)$$

A similarity of the reaction pattern has been shown in the oxidation of alkanecarboxylate and alkaneboronate anions using a platinum anode. The formation of alkyl radicals and cations in the oxidation of alkane-carboxylic acids has well been known as the Kolbe and abnormal Kolbe reactions. Almost the same products have been obtained in the oxidation of the corresponding alkaneboric acids (22) [17].

$$RCO_2^- \xrightarrow{-e} RCO_2\cdot \longrightarrow R\cdot + CO_2$$
$$\xrightarrow{-e} R^+$$
$$R\bar{B}(OH)_3 \xrightarrow{-e} R\dot{B}(OH)_3 \longrightarrow R\cdot + B(OH)_3$$

$$(22)$$

Table 2. Coupling of Radicals

R in R_3B	R−R	Yield (%)
C_8H_{17}	$C_{16}H_{34}$	69
C_6H_{13}	$C_{12}H_{26}$	56
C_5H_{11}	$C_{10}H_{22}$	46
cyclo-C_6H_{11}	$C_6H_{11}-C_6H_{11}$	23

The generation of alkyl radicals and cations by the anodic oxidation of alkaneboronate anions has been utilized for the anodic oxidation of trialkylboranes in the presence of hydroxide or alkoxide anions (23), (Table 2) [18–20].

$$R_3B \ + \ OY^- \ \longrightarrow \ R_3\bar{B}OY \ \xrightarrow{-e} \ R_3\dot{B}OY \ \longrightarrow \ R \cdot \ + \ R_2BOY$$

$$R_2BOY \ + \ OY^- \ \longrightarrow \ R_2\bar{B}(OY)_2 \ \xrightarrow{-e} \ R_2\dot{B}(OY)_2 \ \longrightarrow \ R \cdot + RB(OY)_2$$

$$RB(OY)_2 \ + \ OY^- \ \longrightarrow \ R\bar{B}(OY)_3 \ \xrightarrow{-e} \ R\dot{B}(OY)_3 \ \longrightarrow \ R \cdot \ + \ B(OY)_3$$
(23)

$$R \cdot + R \cdot \ \longrightarrow \ R_2 \qquad R \cdot \xrightarrow[\substack{carbon \\ anode}]{-e} R^+ \xrightarrow{OY^-} ROY$$

$$Y : H \ or \ CH_3$$

The alkyl radicals generated according to scheme (23) can be trapped by butadiene [18].

The anodic oxidation of trialkylboranes in the presence of bromides or iodides as supporting electrolytes shows different reaction patterns involving the formation of alkyl halides from trialkylboranes and halogen radicals generated by the anodic oxidation of the halide anions.

$$X^- \ \xrightarrow{-e} \ X \cdot$$
(24)

$$R_3B \ + \ X \cdot \ \longrightarrow \ R_2B \cdot \ + \ RX$$

The type of reaction shown by Eq. (24) is described in Section 2.5.2.

References

1. Mango, F., Bontempelli, G., Dilloni, G.: J. Electroanal. Chem. *30*, 375 (1971)
2. Bontempelli, G., Mango, F., Mazzocchin, G. A.: J. Electroanal. Chem. *42*, 57 (1973)
3. Bewick, A., Coe, D. E., Mellor, J. M., Walton, D. J.: J. Chem. Soc., Chem. Commun. *1980*, 51
4. a) Houghton, D. S., Humffray, A. A.: Electrochim. Acta *17*, 1421 (1972)
   b) Humffray, A. A., Houghton, D. S.: Electrochim. Acta *17*, 1435 (1972)
5. Cottrell, P. T., Mann, C. K.: J. Electrochem. Soc. *116*, 1499 (1969)
6. a) Glass, R. S., Duchek, J. R., Klug, J. T., Wilson, G. S.: J. Am. Chem. Soc. *99*, 7349 (1977)
   b) See also Wilson, G. S., Swanson, D. D., Klug, J. T., Glass, R. S., Ryan, M. D., Musker, W. K.: J. Am. Chem. Soc. *101*, 1040 (1979)
7. a) Mango, F., Bontempelli, G.: J. Electroanal. Chem. *36*, 389 (1972)
   See also b) Uneyama, K., Torii, S.: J. Org. Chem. *37*, 367 (1972)
   c) Torii, S., Matsuyama, Y., Kawasaki, K., Uneyama, K.: Bull. Chem. Soc. Jpn. *46*, 2912 (1973)

8. Hoffelner, H., Yorgiyadi, S., Wendt, H.: J. Electroanal. Chem. *66*, 138 (1975)
9. Shono, T., Matsumura, Y., Hayashi, J., Usui, M.: Denki Kagaku *51*, 131 (1983)
10. Canfield, N. D., Chambers, J. Q.: J. Electroanal. Chem. *56*, 459 (1974)
11. a) Gourcy, J., Martigny, P., Simonet, J., Jeminet, G.: Tetrahedron *37*, 1495 (1981)
    b) See also Porter, O. N., Utley, J. H. P.: J. Chem. Soc., Chem. Commun. *1978*, 255
12. Bontempelli, G., Magno, F., Mazzocchin, G. A.: J. Electroanal. Chem. *55*, 109 (1974)
13. Blankespoor, R. L., Doyle, M. P., Hedstrand, D. M., Tamblyn, W. H., Van Dyke, D. A.: J. Am. Chem. Soc. *103*, 7096 (1981)
14. Ohmori, H., Nakai, S., Masui, M.: J. Chem. Soc., Perkin Trans. (1) *1979*, 2023
15. Bancroft, E. E., Blount, H. N., Janzen, E. G.: J. Am. Chem. Soc. *101*, 3692 (1979)
16. a) Geske, D. H.: J. Phys. Chem. *66*, 1743 (1962)
    b) See also Turner, W. R., Elving, P. J.: Anal. Chem. *37*, 207 (1965)
17. Keating, J. T., Skell, P. S.: J. Org. Chem. *34*, 1479 (1969)
18. Schäfer, H. J., Koch, D.: Angew. Chem. Int. Ed. Engl. *11*, 48 (1972)
19. Taguchi, T., Itoh, M., Suzuki, A.: Chem. Lett. *1973*, 719
20. Taguchi, T., Takahashi, Y., Itoh, M., Suzuki, A.: Chem. Lett. *1974*, 1021

2.5 Oxidation of Organic Halides and Oxidative Halogenation of Organic Compounds

The electroorganic reactions discussed in this section are classified into the following three categories:

(a) Oxidation of alkyl or aryl halides generating carbonium ion intermediates (1).

(b) Oxidation of halide anions followed by the reaction of the resulting halogen radicals or cations with organic compounds (2).

(c) Reactions in which the cationic halogen species behaves as mediators.

$$RX \xrightarrow{-e} [RX]^{+\cdot} \longrightarrow R^+ + X\cdot \tag{1}$$

$$X^- \xrightarrow{-2e} X^+ \xrightarrow{Nu^-} NuX \qquad Nu^- : \text{nucleophilic compounds} \tag{2}$$

The third type of reactions is described in the Section 2.9.3.

2.5.1 Oxidation of Organic Halides

The oxidation potential required for the removal of one electron from the nonbonding orbital of the halogen of alkyl iodides ($E_{1/2}$, 1.9–2.1 V *vs.* Ag/Ag$^+$) and bromides (E_p, 2.5–2.8 V *vs.* Ag/Ag$^+$) is in the range accessible by the electrochemical method whereas oxidation of alkyl chlorides and fluorides is hardly achievable.

The anodic oxidation of alkyl iodides [1] or bromides [2] in acetonitrile yields the corresponding N-alkylacetamides suggesting the intermediate formation of carbonium ions or their equivalents (3).

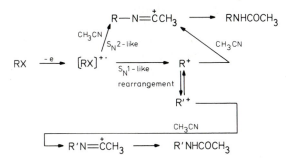

$$ (3) $$

The carbonium ions have been shown to be similar to those formed in the solvolysis of the corresponding alkyl tosylates in trifluoroacetic acid rather than those formed in the abnormal Kolbe electrolysis of the corresponding carboxylic acids [3]. The relative rates of S_N1-like and S_N2-like routes and the rearrangement of the intermediate carbonium ions depend on the structure of the groups R since the nucleophilicity of acetonitrile is not high enough to make the S_N2-like route dominant.

In the oxidation of α,α-dideutero-β-phenylethyl iodide, for example, deuterium is completely distributed in the generated products whereas the oxidation of α,α-dideuteropropyl iodide gives only the products where deuterium is not rearranged (4).

$$ (4) $$

The fact that the extent of racemization observed in the anodic oxidation of optically active α-phenylethyl iodide is 75% suggests the participation of S_N2-like process [4]. A rearrangement characteristic of the carbonium ions involving a strained ring system has been observed in the oxidation of cyclopropyl bromide (5) [5] and iodomethylcyclopropane (6) [6].

$$ (5) $$

$$ (6) $$

major (>90 %)

The anodic acetamidation of polycyclic bridgehead iodide proceeds smoothly for compounds of low steric strain whereas highly strained compounds yield complex mixtures of products suggesting electron transfer from the cage-ring system rather than from the nonbonding orbital of iodine [7]. Adamantyl fluoride, chloride, and bromide are exceptional compounds, where the electron is transferred from the adamantyl moiety to anode, and a single product is obtained in high yield from each halide [8].

In contrast to alkyl iodides, the carbon-iodine bond of aryl iodides is not cleaved in the anodic oxidation, but coupling takes place. Thus, the major product obtained in the oxidation of iodobenzene ($LiClO_4$, CH_3CN) is 4-iododiphenyliodonium perchlorate [1].

The formation of this salt may be explained by the mechanism according to which an initially formed iodobenzene cation radical attacks iodobenzene yielding a second cation radical which decomposes to the oxidation product by loss of one proton and one electron (7).

$$ArI \xrightarrow{-e} ArI^{+\cdot} \xrightarrow{C_6H_5I} \quad \text{(intermediate)} \xrightarrow{-e} Ar-\overset{+}{I} \text{—}\bigcirc\text{—} I \;+\; H^+ \qquad (7)$$

2.5.2 Halogenation

The cationic species formed by the anodic oxidation of halide anions add to alkenes in the presence of suitable nucleophiles (8).

$$\overset{\diagdown}{\diagup}C=C\overset{\diagup}{\diagdown} \;+\; X^- \xrightarrow[Nu]{-2e} \overset{\diagdown}{\diagup}\underset{Nu}{C}-\underset{X}{C}\overset{\diagup}{\diagdown} \qquad (8)$$

Early investigations on the oxidative addition of halogen to alkenes have mainly been focused on the preparation of epoxides from lower alkenes such as ethylene and propylene [9]. The anodic halo-methoxylation of cinnamic acid is probably one of the earliest studies dealing with higher alkenes and related compounds [10]. Table 1 summarizes some of the typical results of the oxidative addition of halogen to higher alkenes.

The positively charged species of halogen also add to vinyl ethers and their derivatives [18].

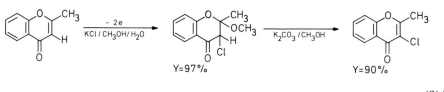

$$Y=97\% \qquad\qquad Y=90\%$$

$$(9)\,[19]$$

Table 1. Oxidative Addition of Halogen

Alkene	Solvent	Supporting electrolyte	Product	Yield (%)	Ref.	Year
$C_6H_5CH=CHCO_2H$	CH_3OH	NH_4Cl	$C_6H_5CHCHCO_2H$ (Cl; OCH$_3$)	50	[10]	1945
trans-$C_6H_5CH=CHC_6H_5$	CH_3OH	NH_4Br	$C_6H_5-CH-CH-C_6H_5$ (Br, Br)	30	[11]	1967
			$C_6H_5-CH-CH-C_6H_5$ (Br; OCH$_3$)	17		
[cyclohexene]	CH_3CN	$(C_2H_5)_4NCl$	[cyclohexane ring with Cl and NHCOCH$_3$]	80	[12]	1970
[cyclohexene]	CH_3CN	I_2	[cyclohexane ring with NHCOCH$_3$ and I]	~100	[13]	1971
$C_6H_5CH_2CH=CHCH_3$	CH_3OH	$(C_2H_5)_4NBr$	$C_6H_5CH_2CH-CHCH_3$ (Br, Br)	35	[13]	1971
			$C_6H_5CH_2CHCHCH_3$ (OCH$_3$; Br)	50		
			$C_6H_5CH_2CHCHCH_3$ (OCH$_3$; OCH$_3$)	27		
[cyclohexene]	CH_3CO_2H	$(C_2H_5)_4NBr$	[cyclohexane ring with OCOCH$_3$ and Br]	82	[14]	1972
$C_6H_5CH=CHCH_3$	DMF	$(C_2H_5)_4NCl$	$C_6H_5CHCHCl-CH_3$ (OCHO)	85	[15]	1978
$C_6H_5C\equiv CC_6H_5$	DMF	$(C_2H_5)_4NCl$	$C_6H_5-C=CClC_6H_5$ (OCHO)	84	[15]	1978
[cyclohexene]	DMF	$(C_2H_5)_4NCl$	[cyclohexane ring with Cl and OCHO]	57	[16]	1979 1980
			[cyclohexane ring with Cl and Cl]	22		
[cyclohexene ring with CO$_2$CH$_3$, CO$_2$CH$_3$]	CH_3OH	HBr	[cyclohexane ring with CH$_3$O, CO$_2$CH$_3$, Br, CO$_2$CH$_3$]	88	[17]	1981

2. Anodic Oxidations

Halogenation of aromatic nuclei may also be achieved by halogen or positively charged species of halogen formed in solution by anodic oxidation of halide anions (10).

$$ArH + X^- \xrightarrow{-2e} ArX + H^+ \qquad (10)$$

Since in earlier investigations the reaction mechanism has not always been elucidated, only some of the recent typical results are summarized in Table 2.

Table 2. Halogenation of Aromatic Compounds

Aromatic compound	Solvent	Halide	Product	Yield (%)	Ref.
Toluene	CH_3CO_2H	LiCl	o- and p-Chlorotoluene	90	[20]
Benzene	CH_3CN	$LiCl/LiClO_4$	Chlorobenzene	90	[21]
Benzene	CH_3CN	$AlCl_3/LiClO_4$	Chlorobenzene	80	[21]
p-Xylene (with light)	CH_3CO_2H	NaBr	Bromo-p-xylene	20	[22]
Naphthalene (with light)	CH_3CO_2H	NaBr	Bromonaphthalene	65	[23]
Phenol	0.5 M $HClO_4$	NaBr	p-Bromophenol	52	[24]
Xylene	CH_3CN	$I_2-LiClO_4$	Iodo-p-xylene	50	[25]
Benzonitrile	$ClCH_2CH_2Cl$ + CF_3CO_2H (10%)	$I_2-(C_4H_9)_4NBF_4$	m-Iodobenzonitrile	40	[26]
Nitrobenzene	$ClCH_2CH_2Cl$ + CF_3CO_2H (10%)	$I_2-(C_4H_9)_4NBF_4$	m-Iodonitrobenzene	78	[26]

The anodic oxidation of a mixture of iodine and aromatic compounds in acetonitrile gives aryl iodides in rather low yields. The iodination is highly improved by using a stepwise method (11) [27].

$$I_2 \xrightarrow[CH_3CN]{-2e} CH_3\overset{+}{C}{=}NI \xrightarrow[ArH]{} ArI + CH_3CN + H^+ \qquad (11)$$

Thus, in a typical experiment, the oxidation of iodine is performed at a platinum anode in acetonitrile containing lithium perchlorate. A stable, pale-yellow solution is obtained after about 2 F/mol of electricity has passed through the solution of iodine. After the oxidation is terminated, a tenfold excess of the aromatic compound is added to the solution. Some of the results are listed in Table 3.

Table 3. Iodination

Aromatic compound	Yield (%)	Isomers (%)			Aromatic compound	Yield (%)	Isomers (%)		
		o-	m-	p-			o-	m-	p-
Anisole	90	33	10	57	Bromobenzene	60	24		76
Toluene	93	48	4	48	Iodobenzene	70	33		67
Benzene	91				Phenyl acetate	55	36		64
Fluorobenzene	85	10		90	Ethylbenzene	65		100	
Chlorobenzene	80	20		80	p-Xylene	90			

Anodic formation of chlorine from chloride anions in solution has been used for regioselective chlorination of a steroid [28].

It has been shown that the reaction of trialkylboranes with iodine radicals generated by the anodic oxidation of iodide anions gives the cor-

Table 4

Substrate	Supporting electrolyte	Product	Yield (%)	Ref.
RR′CHCH₂OH	. HCl	RR′CCH(OCH₂CHRR′)₂ / Cl	61 – 83	[34]
C₂H₅OH, CH₃COCH₃	NaI	CHI₃	98	[35] [36]
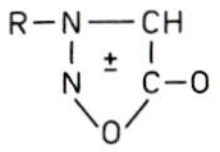 R—N—CH (ring) N—C—O	NaBr, (CH₃)₄NCl	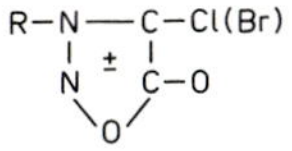 R—N—C—Cl(Br) (ring) N—C—O	90 – 100	[37]
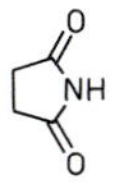 succinimide NH	NaBr	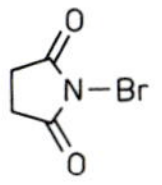 N—Br	66	[38] [41] [43]
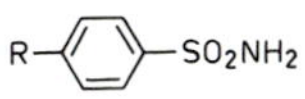 R—C₆H₄—SO₂NH₂	NaCl	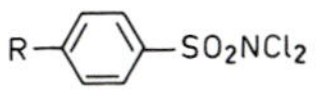 R—C₆H₄—SO₂NCl₂	80 – 90	[39]
R(CONH₂)CONH₂	NaCl	R(CONHCl)CONHCl	80 – 90	[40]
RNHCO₂C₂H₅	NaCl, NaBr	R—NCO₂C₂H₅ / Cl(Br)	40 – 60	[42]
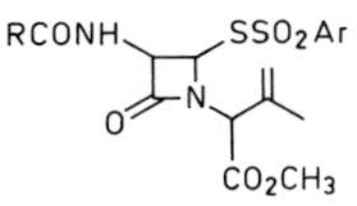 RCONH—SSO₂Ar ... CO₂CH₃	NaCl	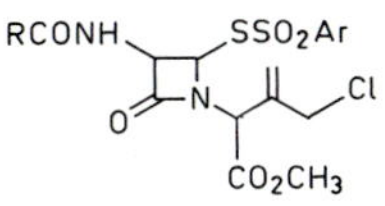 RCONH—SSO₂Ar ...Cl CO₂CH₃	83	[44]

R = C₆H₅CH₂
Ar = −C₆H₄−NO₂−(p)

51

responding alkyl iodides which can be used for the alkylation of acetonitrile [29], nitromethane [30], acetylenes [31–32], and ethyl acrylate [33].

$$R_3B + CH_3CN + I^- \xrightarrow{-[e]} RCH_2CN$$

$$R_3B + CH_3NO_2 + I^- \xrightarrow{-[e]} RCH_2NO_2$$

$$(12)-(15)$$

$$R_3B + C_6H_5C{\equiv}CH + I^- \xrightarrow{-[e]} C_6H_5C{\equiv}CR$$

$$R_3B + CH_2{=}CHCO_2C_2H_5 + I^- \xrightarrow{-[e]} RCH_2CH_2CO_2C_2H_5$$

The electrogenerated halogen has been used for the preparation of a variety of compounds some of which are compiled in Table 4.

References

1. Miller, L. L., Hoffmann, A. K.: J. Am. Chem. Soc. *89*, 593 (1967)
2. a) Becker, J. Y., Münster, M.: Tetrahedron Lett. *1977*, 455
   b) Becker, J. Y.: J. Org. Chem. *42*, 3997 (1977)
3. a) Laurent, A., Laurent, E., Tardivel, R.: Tetrahedron Lett. *1973*, 4861
   b) Laurent, A., Laurent, E., Tardivel, R.: Tetrahedron *30*, 3423 (1974)
4. Laurent, A., Laurent, E., Tardivel, R.: Tetrahedron *30*, 3431 (1974)
5. Becker, J. Y., Zemach, D.: J. Chem. Soc., Perkin Trans. (2) *1974*, 914
6. Laurent, E., Tardivel, R.: Tetrahedron Lett. *1976*, 2779
7. Abeywickrema, R. S., Della, E. W., Fletcher, S.: Electrochim. Acta *27*, 343 (1982)
8. a) Koch, V. R., Miller, L. L.: J. Am. Chem. Soc. *95*, 8631 (1973)
   b) Koch, V. R., Miller, L. L.: Tetrahedron Lett. *1973*, 693
   See also c) Vincent, F., Tradivel, R., Mison, P.: Tetrahedron Lett. *1975*, 603
   d) Vincent, F., Tradivel, R., Mison, P.: Tetrahedron *32*, 1681 (1976)
9. Dietz, R., Lund, H. in: Organic Eledtrochemistry (ed.) Baizer, M. M., p. 821, New York, Marcel Dekker 1973
10. Leininger, R., Pasiut, L. A.: Trans. Electrochem. Soc. *88*, 73 (1945)
11. Inoue, T., Koyama, K., Matsuoka, T., Tsutsumi, S.: Bull. Chem. Soc. Jpn. *40*, 162 (1967)
12. Faita, G., Fleischmann, M., Pletcher, D.: J. Electroanal. Chem. *25*, 455 (1970)
13. Weinberg, N. L., Hoffman, K.: Can. J. Chem. *49*, 740 (1971)
14. Shono, T., Ikeda, A.: J. Am. Chem. Soc. *94*, 7892 (1972)
15. Verniette, M., Daremon, C., Simonet, J.: Electrochim. Acta *23*, 929 (1978)
16. a) Pouillen, P., Minko, R., Verniette, M., Martinet, P.: Electrochim. Acta *25*, 711 (1980)
    b) Pouillen, P., Minko, R., Verniette, M., Martinet, P.: Electrochim. Acta *24*, 1189 (1979)

17. Torii, S., Uneyama, K., Tanaka, H., Yamanaka, T., Yasuda, T., Ono, M., Kohmoto, Y.: J. Org. Chem. *46*, 3312 (1981)
18. Torii, S., Inokuchi, T., Misima, S., Kobayashi, T.: J. Org. Chem. *45*, 2731 (1980)
19. Yamauchi, M., Katayama, S., Nakashita, Y., Watanabe, T.: Synthesis *1981*, 33
20. a) Mastragostino, M., Gasalbore, G., Valcher, S.: J. Electroanal. Chem. *56*, 117 (1974)
    b) See also Mastragostino, M., Gasalbore, G., Valcher, S.: J. Electroanal. Chem. *44*, 37 (1973)
21. Gourcy, J., Simonet, J., Jaccaud, M.: Electrochim. Acta *24*, 1039 (1979)
22. Casalbore, G., Mastragostino, M., Valcher, S.: J. Electroanal. Chem. *61*, 33 (1975)
23. Casalbore, G., Mastragostino, M., Valcher, S.: J. Electroanal. Chem. *68*, 123 (1976)
24. Bejerano, T., Gileadi, E.: Electrochim. Acta *21*, 231 (1976)
25. Miller, L. L., Kujawa, E. P., Campbell, C. B.: J. Am. Chem. Soc. *92*, 2821 (1970)
26. Lines, R., Parker, V. D.: Acta Chim. Scand. *B34*, 47 (1980)
27. Miller, L. L., Watkins, B. F.: J. Am. Chem. Soc. *98*, 1515 (1976)
28. Breslow, R., Goodin, R.: Tetrahedron Lett. *1976*, 2675
29. Takahashi, Y., Tokuda, M., Itoh, M., Suzuki, A.: Chem. Lett. *1975*, 523
30. Takahashi, Y., Tokuda, M., Itoh, M., Suzuki, A.: Synthesis, *1976*, 616
31. Takahashi, Y., Tokuda, M., Itoh, M., Suzuki, A.: Chem. Lett. *1977*, 999
32. Takahashi, Y., Tokuda, M., Itoh, M., Suzuki, A.: Chem. Lett. *1980*, 461
33. a) Takahashi, Y., Yuasa, K., Tokuda, M., Suzuki, A.: Bull. Chem. Soc. Jpn. *51*, 339 (1978)
    b) See also Takahashi, Y., Yuasa, K., Tokuda, M., Itoh, M., Suzuki, A.: Chem. Lett. *1978*, 669
34. White, D. A., Coleman, J. P.: J. Electrochem. Soc. *125*, 1401 (1978)
35. Teeple, J. E.: J. Am. Chem. Soc. *26*, 170 (1904)
36. Ramaswamy, R., Venkatachalapathyl, M. S., Udupa, H. V. K.: J. Electrochem. Soc. *110*, 294 (1963)
37. Tien, Hsien-Ju, Nonaka, T., Sekine, T.: Chem. Lett. *1979*, 283
38. Lamchen, M.: J. Chem. Soc. *1950*, 747
39. a) Miyazaki, H.: Denki Kagaku *44*, 409 (1976)
    b) Miyazaka, H.: Denki Kagaku *45*, 244 (1977)
40. Miyazaki, H.: Denki Kagaku *45*, 475 (1977)
41. Miyazaki, H.: Denki Kagaku *45*, 553 (1977)
42. Miyazaki, H.: Denki Kagaku *46*, 270 (1978)
43. Miyazaki, H.: Denki Kagaku *48*, 453 (1980)
44. a) Torii, S., Tanaka, H., Saitoh, N., Siroi, T., Sasaoka, M., Nokami, J.: Tetrahedron Lett. *23*, 2187 (1982)
    b) See also Balsamo, A., Benedini, P. M., Giorgi, I., Macchia, B., Macchia, F.: Tetrahedron Lett. *23*, 2991 (1982)

2.6 Oxidation of Aliphatic and Aromatic Amines

2.6.1 Oxidation of Aliphatic Amines

The relatively low oxidation potentials of simple aliphatic amines indicate [1][1] that they are easily oxidized by the anodic method. Hence, the anodic oxidation of amines has extensively been studied though many of these studies have been carried out mainly from the viewpoint of the mechanism of electrochemical reaction. Since this book is concerned with electroorganic reactions from the standpoint of organic synthesis, the mechanism of oxidation of amines is surveyed only briefly.

In the presence of an adequate amount of water, aliphatic amines are generally dealkylated in the anodic oxidation [2]. Thus, a tertiary amine is successively dealkylated to a secondary amine, a primary amine, and finally to ammonia. The mechanism involves initial removal of one electron from the lone-pair electrons of nitrogen leading to a cation radical as exemplified by Eq. (1), though a variety of mechanisms have been proposed, depending on the reaction conditions.

$$(RCH_2)_3N \xrightarrow{-e} (RCH_2)_3\overset{+\cdot}{N} \xrightarrow[-H^+]{} (RCH_2)_2N\dot{C}HR \xrightarrow{-e} (RCH_2)_2N\overset{+}{C}HR \rightleftharpoons$$

$$(RCH_2)_2\overset{+}{N}=CHR \xrightarrow[H_2O]{-H^+} (RCH_2)_2NH \quad + \quad RCHO \tag{1}$$

When an anodically oxidizable electrode material such as silver [3] or nickel [4] is used, the reaction pathways are different as illustrated by scheme (2).

$$RCH_2NH_2 \xrightarrow{-e} RCH_2\overset{+\cdot}{NH_2} \xrightarrow[-H^+]{} RCH_2\dot{NH} \xrightarrow{-e} RCH_2\overset{+}{NH} \xrightarrow[-H^+]{} RCH=NH \xrightarrow{-e}$$

$$RCH=\overset{+\cdot}{NH} \xrightarrow[-H^+]{} RCH=\dot{N} \xrightarrow{-e} RCH=\overset{+}{N} \xrightarrow[-H^+]{} RCN \tag{2}$$

Although the anodic dealkylation of aliphatic amines is not always useful in organic synthesis, it is highly effective to simulate the enzymatic dealkylation of amines and amides catalyzed by the cytochrome P450 monooxygenase system. The remarkable similarity between anodic and enzymatic dealkylation supports that the enzymatic dealkylation involves an electron transfer mechanism similar to that involved in the anodic oxidation [5]. As a typical example (scheme 3), the electrochemical dealkylation has been successfully applied to the synthesis of N-dealkylated metabolites of complex drugs

---

[1] $(C_3H_7)_2NH$, 1.26 V vs. NHE; $(C_3H_7)_3N$, 1.02 V vs. NHE.

whereas the microsomal *N*-dealkylation is only achieved on a very small scale under highly delicate reaction conditions [6].

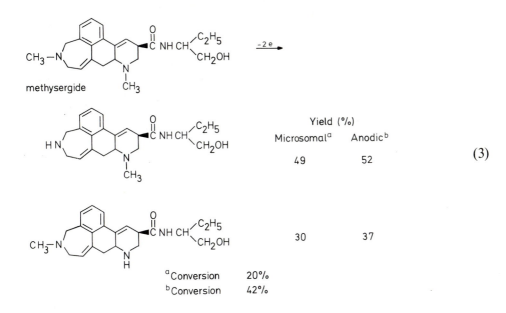

	Yield (%)	
	Microsomal[a]	Anodic[b]
	49	52
	30	37

$$(3)$$

[a] Conversion   20%
[b] Conversion   42%

The reaction pattern of the anodic oxidation of amines greatly depends on the structure of the amine and the reaction conditions including the nucleophilicity of solvents. The anodic oxidation of *N,N*-dimethylbenzylamine in methanol containing tetrabutylammonium fluoroborate or potassium hydroxide affords two substitution products according to a reaction sequence which is different from the oxidative dealkylation (4) [7].

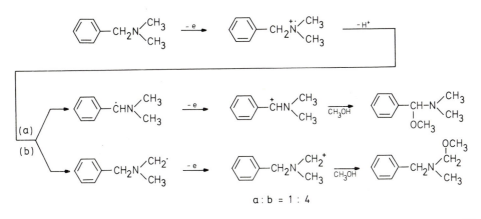

a : b = 1 : 4

$$(4)$$

2. Anodic Oxidations

In the presence of canide anions, the anodic oxidation of tertiary amines yields products cyanated in the position α to the nitrogen atom (Table 1) [8]. Acyclic t-amines bearing an N-methyl group are mainly cyanated at the methyl group whereas in the cyanation of N-methyl substituted alicyclic amines, the substitution mainly takes at the α-carbon in the ring rather than at the N-methyl group. The mechanism of cyanation is almost the same as described by scheme (1) except that the iminium cation intermediate reacts with the cyanide anion rather than with water because the cyanide anion is sufficiently nucleophilic.

The regioselectivity observed in this cyanation has not clearly been explained, but it has been proposed that the conformation of the amine adsorbed on the anode plays an important role in the determination of the regio-selectivity.

The iminium cation intermediate formed by the anodic oxidation of tribenzylamine in acetonitrile can be trapped in $situ$ by diethyl malonate or diethyl phosphite [9].

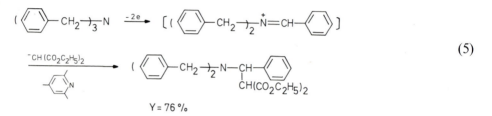

$$(5)$$

In contrast to the instability of cation radicals generated from simple aliphatic amines, it has been observed that cation radicals formed from tetraalkylhydrazines and some diamines having a cage structure show rela-tively long lifetimes [10].

In some special cases, hydrazine derivatives or azo compounds are formed in the anodic oxidation of amines (Table 2).

Tertiary amines possessing special structures like N-benzylaziridine show a different reaction patterns in the anodic oxidation. The formation of a tetramer from N-benzylaziridine has been explained by a unique chain mechanism in which the formation of the iminium cation intermediate is not involved (6) [15].

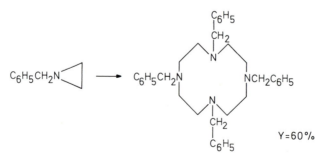

$$(6)$$

Table 1. Anodic Cyanation of Tertiary Amines[a]

Amine	Anode potential (V vs. SCE)	Product	Yield (%)[b]
$(CH_3)_3N$	1.2 – 1.4	$(CH_3)_2NCH_2CN$	53
$(C_2H_5)_3N$	1.1 – 1.2	$(C_2H_5)_2NCHCN$ $\quad\quad\quad\quad$ | $\quad\quad\quad\quad CH_3$	53
$(C_2H_5)_2NCH_3$	1.0 – 1.2	$(C_2H_5)_2NCH_2CN$ + $\begin{array}{c}C_2H_5 \\ \diagdown \\ CH_3 \diagup \end{array}N{-}CHCN$ $\begin{array}{c}\\ \| \\ CH_3\end{array}$ (57) $\quad\quad\quad\quad\quad$ (43)	37
$(i\text{-}C_3H_7)_2NCH_3$	0.9 – 1.1	$(i\text{-}C_3H_7)_2NCH_2CN$	55
☐N–CH₃	1.0 – 1.3	☐N–CH₃ + ☐N–CH₂CN $\quad\quad$ CN $\quad\quad$(81) $\quad\quad\quad$ (19)	59
⬡N–CH₃	1.1 – 1.4	⬡N–CH₃ + ⬡N–CH₂CN $\quad\quad$ CN $\quad\quad$(62) $\quad\quad\quad$ (38)	66
⬡N–C₃H₇–i	1.0 – 1.2	⬡N–C₃H₇–i $\quad\quad$ CN	62

[a]Anolyte: amine (0.10 mol) + NaCN (0.15 mmol) in 75 ml of $CH_3OH - H_2O$ (1:1); [b]GLC

Table 2. Anodic Formation of Hydrazine Derivatives or Azo Compounds

$$2\ R^1R^2NH \xrightarrow{\ -[e]\ } R^1R^2N{-}NR^1R^2 \text{ or } R^1N{=}NR^1 \ (R^2{=}H)$$

R^1	R^2	Yield (%)	Ref.
$t - C_4H_9$	H	41[a]	[11]
$t - C_4H_9$	H	25[a]	[12]
$t - C_4H_9$	H	33[a]	[12]
$cyclo - C_6H_{11}$	H	23[a]	[12]
C_4H_9	C_4H_9	38[b]	[12]
HN⬡ (bicyclic)		58[b]	[13]
(diamine structure)		14[b]	[14]

[a]Azo compounds; [b]Hydrazine derivatives

2. Anodic Oxidations

The anodic oxidation of urazoles in acetonitrile yields 1,2,4-triazoline-3,5-diones which are extremely reactive dienophiles (7) [16].

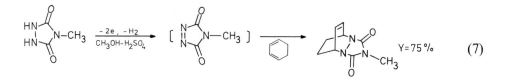

$$Y = 75\% \qquad (7)$$

When benzohydroxamic acid is oxidized anodically in the presence of nucleophiles such as amines or alcohols, benzoylation of the nucleophiles takes place as exemplified below (8) [17].

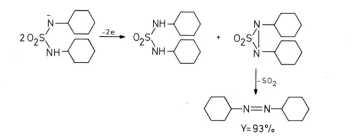

$$(8)$$

The anodic oxidation of N,N-dialkylsulfamides in the presence of base leads to the formation of azoalkanes (9) [18].

$$(9)$$

2.6.2 Oxidation of Aromatic Amines

In contrast to aliphatic amines, the anodic oxidation of aromatic amines shows rather complex reaction patterns which are highly affected by the reaction conditions. Although extensive studies on the electrochemical reaction mechanism have been carried out, examples for the application of the anodic oxidation of aromatic amines to organic synthesis are very few.

The reaction schemes of the oxidation of aniline and its derivatives are summarized in Eq. (10) [19].

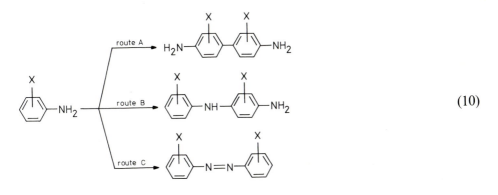

$$(10)$$

The relative rate of each route greatly depends on the reaction conditions and substituents. Thus, route A is rather favored under strongly acidic conditions [20] whereas alkaline conditions facilitate route C [21]. The oxidation of N-alkylanilines and 1-naphthylamine also proceeds in a similar manner as described by scheme (10) [22]. The products in this scheme may be further oxidized and decomposed as exemplified by Eq. (11).

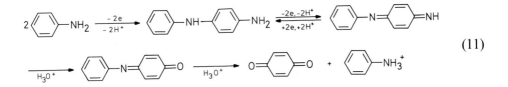

$$(11)$$

Some typical reaction conditions and yields are listed in Table 3 [19].

The anodic oxidation of aromatic amines often gives polymeric products which can be isolated as films [23, 24].

The mechanism of oxidation of diphenylamine and its derivatives has been studied in detail. However, applications to organic synthesis have not been reported so far [25].

Among the anodic reactions of aromatic amines, the anodic cyanation and methoxylation of N,N-dialkylanilines are rather exceptional examples which are useful in organic synthesis. The cyanation takes place at both the phenyl ring and methyl group.

$$(12) \ [26]$$

59

2. Anodic Oxidations

Table 3. Oxidation of Aromatic Amines

Amine	Reaction conditions[a]	Product	Yield (%)[c]
$C_6H_5NH_2$	A, L	$4-NH_2C_6H_4NHC_6H_5$	40
$C_6H_5NH_2$	0.05 M H_2SO_4, H or L	$O=\bigcirc=O$ (PBQ)	90 – 100
$C_6H_5NH_2$	B, H	PBQ	80 – 90
		$NH_2C_6H_4-C_6H_4NH_2$	10 – 20
$C_6H_5NHC_2H_5$	A, H	$C_2H_5NHC_6H_4-C_6H_4NHC_2H_5$	70 – 80
$C_6H_5NHC_2H_5$	A, L[b]	$4-C_2H_5NH-C_6H_4-N-C_6H_5$ $\overset{\textstyle |}{C_2H_5}$	70
		$C_2H_5NHC_6H_4-C_6H_4NHC_2H_5$	15
$C_6H_5NHC_2H_5$	B, H	PBQ	40
		$C_2H_5NHC_6H_4-C_6H_4NHC_2H_5$	60
$C_6H_5NHC_3H_7-iso$	3M H_2SO_4, H	PBQ	35
		$iso-C_3H_7NHC_6H_4-C_6H_4NHC_3H_7-iso$	65
$C_6H_5NHC_4H_9-t$	A, H	$t-C_4H_9NHC_6H_4-C_6H_4NHC_4H_9-t$	100

[a]Controlled potential electrolysis.
[b]Constant-current electrolysis.
A, $CH_3CN-(C_2H_5)_4NClO_4$; B, 6M H_2SO_4; L, low-current density; H, high-current density
[c]Yield based on consumed starting materials.

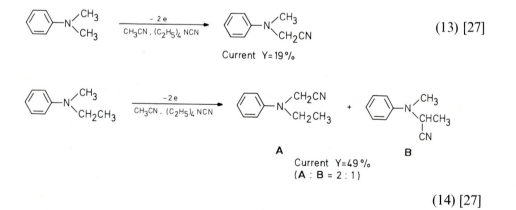

$$\text{Current } Y=19\% \qquad (13) [27]$$

$$\text{A} \qquad \text{B}$$
$$\text{Current } Y=49\%$$
$$(A : B = 2 : 1)$$

$$(14) [27]$$

Methoxylation of *N,N*-dimethyl- or *N*-methyl-*N*-alkylanilines occurs predominantly at the methyl group [28].

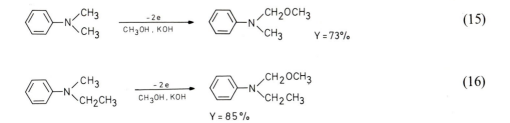

$$\text{(15)}$$

$$Y = 73\%$$

$$\text{(16)}$$

$$Y = 85\%$$

The methoxylated N,N-dimethylaniline is a very useful starting material in organic synthesis since the iminium ion intermediate is easily generated by treatment of the methoxylated compound with Lewis acid. This intermediate can be trapped *in situ* with a variety of nucleophiles such as electron-rich olefins yielding tetrahydroquinolines.

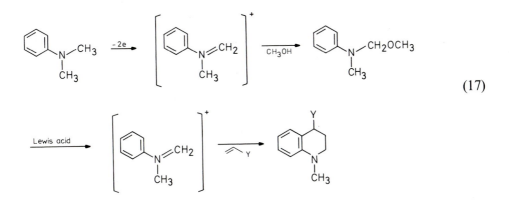

$$\text{(17)}$$

Typical results are summarized in Table 4.

Further anodic oxidation of N-methoxymethyl-N-methylaniline in methanol yields N,N-bis(methoxymethyl)aniline, which is converted into julolidine derivatives upon reaction with two molecules of ethyl vinyl ether.

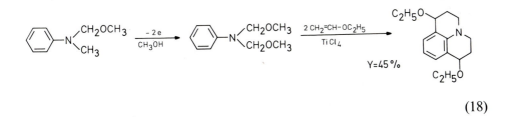

$$Y = 45\%$$

$$\text{(18)}$$

As described above, the reaction of N-methoxymethyl-N-methylaniline with electron-rich olefins gives the corresponding tetrahydroquinoline derivatives. On the other hand, when N-acetoxymethyl-N-methoxycarbonyl-aniline is used as the starting material, quinoline derivatives are obtained

Table 4. Synthesis of Tetrahydroquinolines from N-methoxymethyl-N-methylaniline Using TiCl$_4$ as a Catalyst

Electron-rich olefin	Product	Y^1	Y^2	Y^3	Yield (%)
C$_6$H$_5$CH=CH$_2$	C$_6$H$_5$	H	H		84
C$_6$H$_5$⟩C=CH$_2$ CH$_3$	C$_6$H$_5$	CH$_3$	H		89
C$_6$H$_{13}$CH=CH$_2$	C$_6$H$_{13}$	H	H		58
C$_2$H$_5$OCH=CH$_2$	OC$_2$H$_5$	H	H		64
C$_2$H$_5$CH=CHOSi(CH$_3$)$_3$	OSi(CH$_3$)$_3$	H	C$_2$H$_5$		61
(CH$_3$)$_3$SiO⟩C=CH$_2$ C$_6$H$_5$	OSi(CH$_3$)$_3$	C$_6$H$_5$	H		31
C$_4$H$_9$⟩C=CH$_2$ (CH$_3$)$_3$SiO	OSi(CH$_3$)$_3$	C$_4$H$_9$	H		29
	OCH$_3$	C$_4$H$_9$	H		11
	OH	C$_4$H$_9$	H		21
C$_2$H$_5$CH=CH—N◯O	N◯O	H	C$_2$H$_5$		81
CH$_2$=CHOAc	OCH$_3$	H	H		11
	OH	·H	H		69

in reasonable yields by a similar reaction sequence. The anodic oxidation of carbamates is described in the following section.

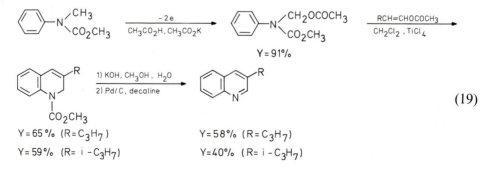

$$Y = 91\%$$

Y = 65 % (R = C$_3$H$_7$)
Y = 59 % (R = i -C$_3$H$_7$)

Y = 58 % (R = C$_3$H$_7$)
Y = 40 % (R = i -C$_3$H$_7$)

(19)

2.6.3 Oxidation of Amides and Carbamates

As described in the previous section, the anodic oxidation of aliphatic amines is utilized only infrequently in organic synthesis, due to the instability of the generated intermediates whereas amides and carbamates of aliphatic amines yield relatively stable intermediates which are sufficiently promising as starting materials in organic synthesis.

The anodic oxidation of ε-caprolactam and N-methyl-ε-caprolactam is the first study which shows that the oxidative transformation of a methyl or a methylene group to a carbonyl group takes place at the position adjacent to the amine nitrogen atom [29]. Since then, a variety of studies have been carried out the results of which are briefly surveyed in Table 5.

The reaction mechanism of the α-methoxylation or α-acetoxylation of amides [46–48] and carbamates [37] has been shown to involve direct one-electron removal from the lone-pair electrons of the nitrogen in the initial step when inert supporting electrolytes are used.

$$RCH_2-\overset{|}{\underset{R'}{N}}-YZ \xrightarrow{-e} RCH_2-\overset{+\cdot}{\underset{R'}{N}}-YZ \xrightarrow[-H^+]{-e} R\overset{+}{C}H-\overset{|}{\underset{R'}{N}}-YZ \xrightarrow{SH} R\underset{S}{C}H-\overset{|}{\underset{R'}{N}}-YZ$$

SH: solvent (CH$_3$OH, CH$_3$CO$_2$H, etc.), Y: CO, PO, or SO$_2$, Z: R″ or OR″

$$(20)$$

The fact that the oxidation potentials of a variety of amides depend on the structures of Y and Z (Table 6) also suggests that the mechanism of direct removal of an electron from the lone-pair electrons on nitrogen [49].

On the other hand, when the oxidation potentials of the supporting electrolytes are lower than those of the amides, the participation of radical species generated by the oxidation of the supporting electrolytes has been suggested [48].

2.6.4 Application of α-Methoxy- or α-Acetoxy-Amides or -Carbamates as Starting Materials in Organic Synthesis

The products obtained by the anodic oxidation of amides or carbamates in methanol have the same structures as the compounds which can be synthesized from amides (carbamates), aldehydes, and methanol. The regeneration of iminium cations from these α-methoxyamides and subsequent reactions of the iminium cations with nucleophiles such as active

2. Anodic Oxidations

Table 5. Anodic Oxidation of Amides, Carbamates, and Lactams

Substrate	Product	Yield (%)	Ref.	Year
(CH$_2$)$_4$(CO)(CH$_2$)N—CH$_3$	(CH$_2$)$_4$(CO)(CH$_2$)NH, CH$_2$O	—	[29]	1954
	(CH$_2$)$_4$(CO$_2$H)(CO$_2$H)	—		
(CH$_3$)$_2$CHNHCOCH$_2$CH$_2$CONH$_2$	(CH$_3$)$_2$C=O	77.5	[30]	1960
	(CH$_2$)$_2$(CO$_2$H)(CO$_2$H)	76.4		
CH$_3$OCNH(CH$_2$)$_4$NHCOCH$_3$	(CH$_2$)$_2$(CO$_2$H)(CO$_2$H)	62.8	[31]	1961
HCON(CH$_3$)(CH$_3$)	HCONCH$_2$OCH$_2$NCHO, (CH$_3$)(CH$_3$)	—	[32]	1964
HCON(CH$_3$)(CH$_3$)	HCON(CH$_2$OCHO)(CH$_3$)	33	[33]	1964 1966
	HCONCH$_2$OCH$_2$NCHO, (CH$_3$)(CH$_3$)	21		
CH$_3$SO$_2$N(CH$_3$)(CH$_3$)	CH$_3$SO$_2$N(CH$_2$OCH$_3$)(CH$_3$)	81	[34]	1972
(2-pyrrolidinone, NH)	(5-OCH$_3$-2-pyrrolidinone, NH)	52	[35]	1972
(N—CHO pyrrolidine)	(NCHO, OCOC$_6$H$_5$)	—	[36]	1973
(O, N—CHO morpholine)	(O, N—CHO, OCOC$_6$H$_5$)	—	[36]	1973
(C$_3$H$_7$)$_2$NCO$_2$CH$_3$	C$_3$H$_7$(C$_2$H$_5$CH)(OCH$_3$)N—CO$_2$CH$_3$	68	[37] [35]	1975 1972
(N—CO$_2$CH$_3$ piperidine)	(N—CO$_2$CH$_3$, OCH$_3$)	72	[37]	1975
(cyclohexyl)(CH$_3$)NCO$_2$CH$_3$	(cyclohexyl)(CH$_3$OCH$_2$)NCO$_2$CH$_3$	78	[37]	

Table 5. (continued)

Substrate	Product	Yield (%)	Ref.	Year
CH$_3$NCH$_2$CH$_2$OH │ CO$_2$CH$_3$	(structure: pyrrolidinone-type ring with N-CO$_2$CH$_3$)	60	[37]	
HCON(CH$_3$)$_2$	HCON(CH$_3$)(CH$_2$P(C$_6$H$_5$)$_3$), ClO$_4^-$	60	[38]	1975
HCON(CH$_3$)$_2$	HCON(CH$_3$)(CH$_2$OAc)	92	[39]	1975
(structure: pyrrolidine N−CHO)	(structure: pyrrolidine N−CHO, OCH$_3$)	97	[40]	1976
(structure: morpholine N−CHO)	(structure: morpholine N−CHO, OCH$_3$)	96	[40]	1976
OHC−N⌐⌐N−CHO	OHC−N⌐⌐N−CHO, OCH$_3$	91	[40]	
(structure: pyrrolidine N−COCH$_3$)	(structure: OCH$_3$, N−COCH$_3$, OCH$_3$)	—	[41]	1978
(structure: indoline N−COCH$_3$)	(structure: indoline OCOCH$_3$, OCOCH$_3$, N−COCH$_3$)	77	[42]	1978
(structure: piperidinone N−CH$_3$)	(structure: piperidinone OH, N−CH$_3$)	64.3	[43]	1979
(structure: C$_2$H$_5$, (CH$_2$)$_n$CO$_2$H piperidinone)	(structure: C$_2$H$_5$, (CH$_2$)$_n$, O−C=O) n=1 n=2	76 58	[44]	1980
(structure: pyrrolidine N−SO$_2$C$_6$H$_4$−CH$_3$-p)	(structure: N−SO$_2$C$_6$H$_4$−CH$_3$-p, OCH$_3$)	79	[45]	1981
(structure: piperidine N−SO$_2$C$_6$H$_4$−CH$_3$-p)	(structure: OCH$_3$, N−SO$_2$C$_6$H$_4$−CH$_3$-p, OCH$_3$)	94	[45]	1981

Table 6. Oxidation Potentials of Carbamates and Amides

Carbamate and amide	Oxidation potential (E ½ vs. SCE)	α–Methoxylated carbamate and amide	Yield (%)
[cyclic]N−CO$_2$CH$_3$	1.96	[cyclic]N−CO$_2$CH$_3$, OCH$_3$	86
[cyclic]N−COCH$_3$	1.88	—	—
[cyclic]N−SO$_2$C$_6$H$_4$CH$_3$	2.14	[cyclic]N−SO$_2$C$_6$H$_4$CH$_3$, OCH$_3$	21.5
[cyclic]N−PO(OC$_2$H$_5$)$_2$	1.96	[cyclic]N−PO(OC$_2$H$_5$)$_2$, OCH$_3$	64
(CH$_3$)$_2$NPO(OC$_2$H$_5$)$_2$	2.02	CH$_3$OCH$_2$, CH$_3$ > NPO(OC$_2$H$_5$)$_2$	72
(CH$_3$)$_2$NSO$_2$CH$_3$	2.07 – 2.08[a]	CH$_3$OCH$_2$, CH$_3$ > NSO$_2$CH$_3$	81[b]

[a] Peak potential (V vs. Ag/Ag$^+$) [34]
[b] Ref. [34]

methylene compounds or nucleophilic aromatic nuclei have been well known under the term amidoalkylation (21) [50].

$$R^1CONHR^2 + R^3CHO + CH_3OH \longrightarrow R^1CON\underset{R^3}{\overset{R^2}{\mid}}CHOCH_3 \longrightarrow R^1CON\underset{R^2}{\overset{+}{=}}CHR^3 \xrightarrow{Nu}$$

$$R^1CON\underset{R^2}{\overset{Nu}{\mid}}CHR^3 \tag{21}$$

In the amidoalkylation, however, the preparation of the starting α-methoxyamides is often difficult since the use of aldehydes higher than formaldehyde is not necessarily successful, and even when formaldehyde is employed, the yields and purity of the α-methoxyamides are not always satisfactory. On the other hand, the anodic α-methoxylation of amides and carbamates generally allows the synthesis of α-methoxyamides (carbamates) which cannot be prepared by the method described by scheme (21).

Since anodically prepared α-formyloxy-N,N-dimethylformamide has successfully been used as an electrophilic reagent [51] and it has been found that the α-methoxylation of the carbamates of a variety of higher aliphatic amines and alicyclic amines [37] can be readily performed, extensive studies have been carried out to utilize the anodically synthesized α-methoxy- or α-acyloxyamides and -carbamates as electrophiles in organic synthesis [52].

2.6.4.1 Formation of Carbon—Carbon Bonds

Aromatic compounds which are sufficiently reactive as nucleophiles can be used for amidoalkylation.

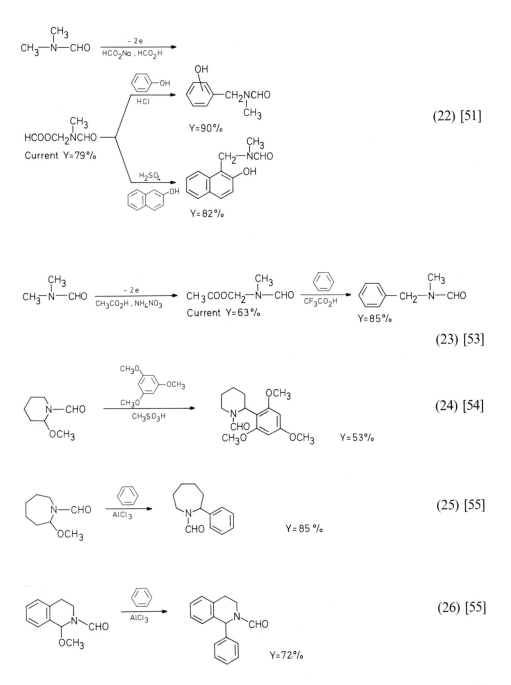

(22) [51]

(23) [53]

(24) [54]

(25) [55]

(26) [55]

67

2. Anodic Oxidations

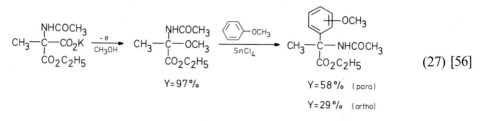

$$Y = 97\%$$

$$Y = 58\% \quad (para)$$
$$Y = 29\% \quad (ortho)$$

(27) [56]

Intramolecular reaction of the iminium cation with an aromatic ring existing in the same compound has also been observed [57].

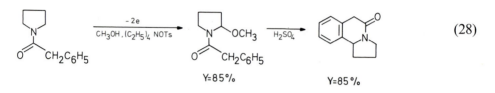

$$Y = 85\%$$

$$Y = 85\%$$

(28)

Heteroaromatic rings such as furan may also be used as nucleophiles in the amidoalkylation.

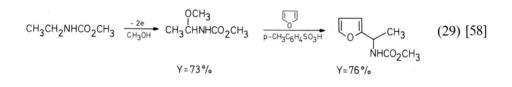

$$Y = 73\%$$

$$Y = 76\%$$

(29) [58]

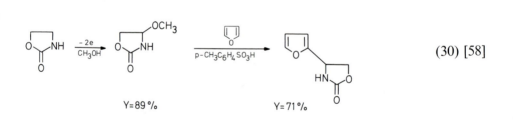

$$Y = 89\%$$

$$Y = 71\%$$

(30) [58]

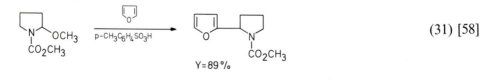

$$Y = 89\%$$

(31) [58]

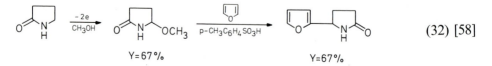

$$Y = 67\%$$

$$Y = 67\%$$

(32) [58]

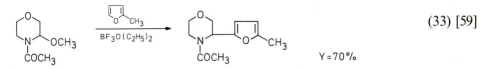

(33) [59]

Y = 70%

The furan ring of the obtained products can be further converted into other skeletons as exemplified by the following three reactions. In the first example, pyridoxine is synthesized in reasonable yield.

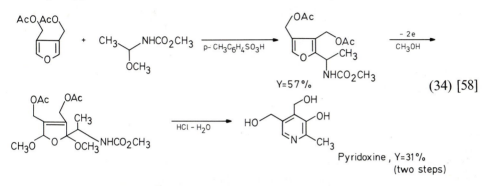

Y=57%

(34) [58]

Pyridoxine, Y=31%
(two steps)

The next two examples clearly show that these reactions are typical and useful methods for the synthesis of 1-azabicyclo[4.n.0]systems.

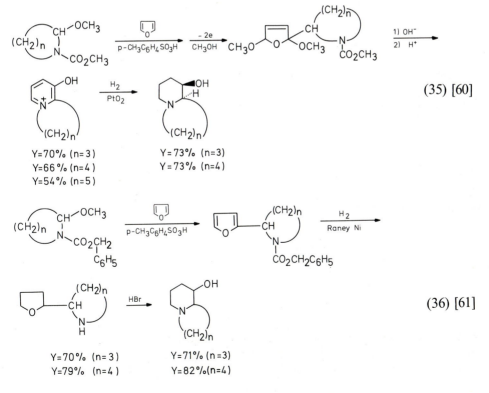

(35) [60]

Y=70% (n=3)
Y=66% (n=4)
Y=54% (n=5)

Y=73% (n=3)
Y=73% (n=4)

(36) [61]

Y=70% (n=3)
Y=79% (n=4)

Y=71% (n=3)
Y=82% (n=4)

2. Anodic Oxidations

Active methylene compounds such as dimethyl malonate may also be used as nucleophiles in the amidoalkylation.

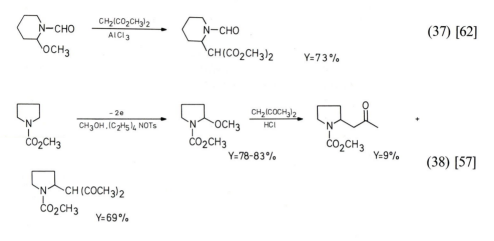

(37) [62]

Y=73%

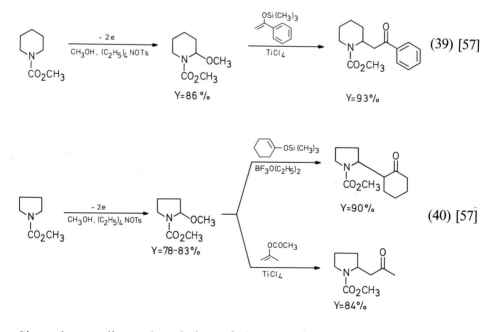

(38) [57]

The nucleophiles unstable under acidic conditions require Lewis acids rather than Brönstead acids as catalysts. Enol ethers or enol esters are typical such nucleophiles.

(39) [57]

(40) [57]

Since the anodic methoxylation of these products takes place at the less substituted α-position, the intramolecular reaction of 2-acetonyl-5-methoxy-1-methoxycarbonylpyrrolidine gives a tropanone-type skeleton.

70

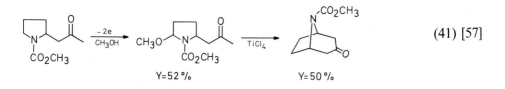

(41) [57]

Y=52%   Y=50%

A cyano group is introduced using trimethylsilylcyanide as the cyanation agent.

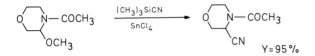

Y=95%

On the other hand, when using isocyanides as nucleophiles, N-alkyl- or N-aryl-aminocarbonyl groups are introduced into the α-position to yield α-amino acid derivatives [63].

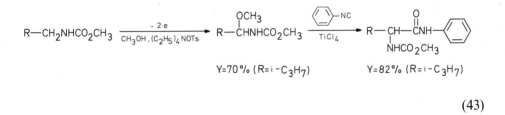

Y=70% (R=i-C_3H_7)   Y=82% (R=i-C_3H_7)

(43)

Although Grignard reagents are the most common nucleophiles, they are highly unstable under acidic conditions. Hence, the treatment of α-methoxylated carbamates with Lewis acid catalysts is necessary before the addition of Grignard reagents to the reaction system. The reaction of Grignard reagents with α-methoxylated carbamates gives poor results without the use of Lewis acid catalysts.

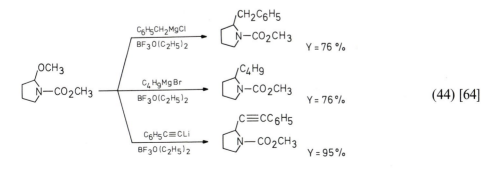

(44) [64]

2. Anodic Oxidations

The reaction of allylmagnesium chloride with α-methoxycarbamates affords only low yields of allylated products whereas high yields are obtained when the Grignard reagent is replaced by allyltrimethylsilane.

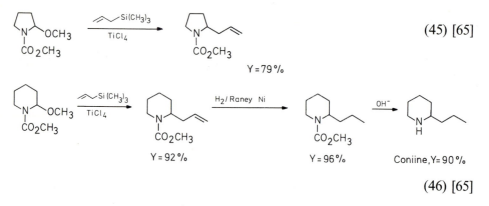

(45) [65]

Y = 79 %

(46) [65]

2.6.4.2 Formation of Bonds between Carbon and Heteroatoms

The reaction of trialkyl phosphites with α-methoxycarbamates leads to the formation of carbon-phosphorous bonds [66].

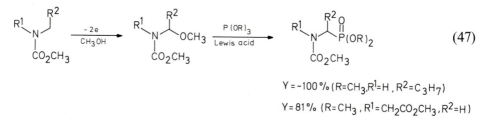

(47)

$Y = ~100 \%$ (R=CH$_3$,R^1=H , R^2=C$_3$H$_7$)

$Y = 81 \%$ (R=CH$_3$, R^1=CH$_2$CO$_2$CH$_3$,R^2=H)

The phosphine oxides prepared by the reaction of α-methoxycarbamates with chlorodiphenylphosphine undergo a variety of reactions in which the α-positions of carbamates can behave as nucleophiles.

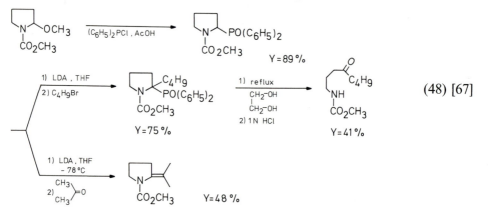

(48) [67]

72

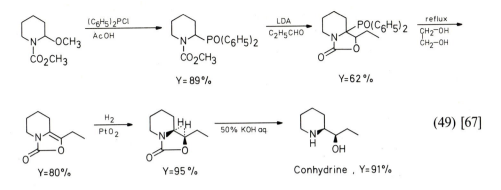

(49) [67]

Conhydrine , Y=91%

The formation of carbon-sulfur bond is exemplified by the following two reactions.

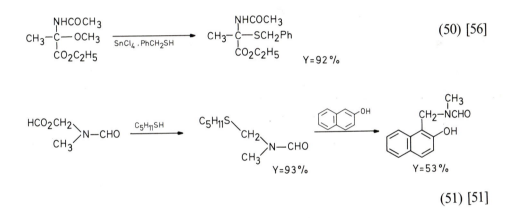

(50) [56]

(51) [51]

The following reactions illustrate the formation of carbon-nitrogen and carbon-oxygen bonds.

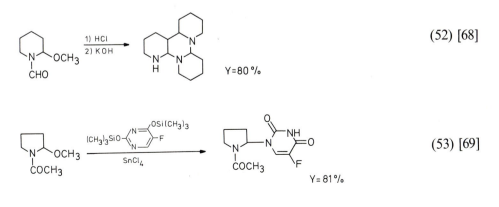

(52) [68]

(53) [69]

73

2. Anodic Oxidations

$$HCO_2CH_2 \diagdown N-CHO \xrightarrow[ROH]{conc.\ HCl} RO-CH_2 \diagdown N-CHO \quad CH_3 \diagup$$

Y = 68 % (R=C_2H_5)
Y = 77% (R= i-C_3H_7)
Y = 88% (R= C_6H_{11})

(54) [51]

2.6.4.3 Reactions Using α-Methoxylated Amides and Carbamates as Masked Carbonyl Compounds

Since α-methoxylated amides and carbamates exhibit the same reactivity as carbonyl compounds, treatment of these α-methoxylated compounds with acidic methanol yields the corresponding acetals.

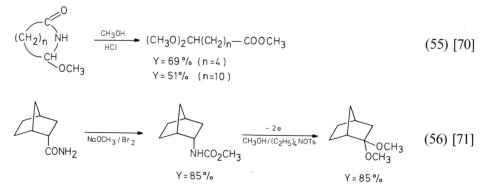

$$\xrightarrow[HCl]{CH_3OH} (CH_3O)_2CH(CH_2)_n-COOCH_3$$

Y = 69 % (n =4)
Y = 51% (n=10)

(55) [70]

$$\xrightarrow{NaOCH_3 / Br_2} \quad \xrightarrow[CH_3OH/(C_2H_5)_4NOTs]{-2e} \quad$$

CONH₂

NHCO₂CH₃

OCH₃

Y = 85%

Y = 85%

(56) [71]

In the following example the α-methoxylated compound is used as a masked aldehyde.

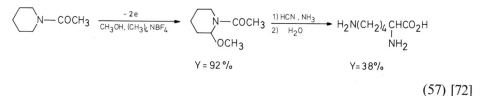

$$\xrightarrow[CH_3OH,\ (CH_3)_4NBF_4]{-2e} \quad \xrightarrow[2)\ H_2O]{1)\ HCN,\ NH_3} H_2N(CH_2)_4\underset{NH_2}{CHCO_2H}$$

Y = 92 %

Y= 38%

(57) [72]

β-Substituted indoles such as indoleacetic acid, tryptamine, and L-tryptophan are easily synthesized utilizing masked aldehydes [73].

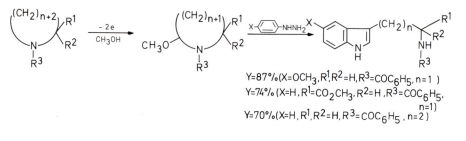

Y=87%(X=OCH₃,R¹,R²=H,R³=COC₆H₅, n=1)
Y=74% (X=H,R¹=CO₂CH₃, R²=H ,R³=COC₆H₅, n=1)
Y=70%(X=H, R¹,R²=H, R³=COC₆H₅ , n=2)

(58)

74

The use of α-methoxylated compounds as masked aldehydes requires acidic catalysts whereas α-hydroxylated amides and carbamates may be employed as masked aldehydes without a catalyst.

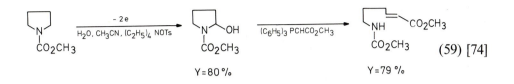

(59) [74]

2.6.4.4 Formation of Enecarbamates and their Reactions

α-Methoxylated carbamates are easily converted into unsaturated carbamates (enecarbamates) through elimination of methanol (Table 7).

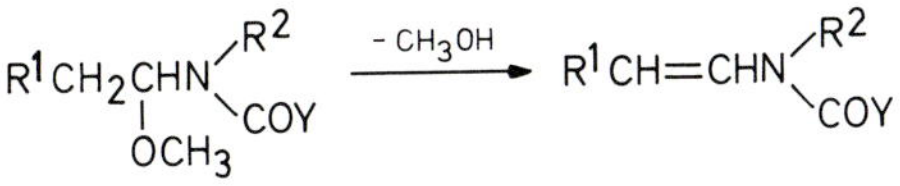

$$R^1CH_2CHN \underset{OCH_3}{\overset{R^2}{\diagdown}} \xrightarrow{-CH_3OH} R^1CH=CHN \overset{R^2}{\underset{COY}{\diagdown}} \tag{60}$$

A variety of substituents are introduced into the α- or β-position of carbamates by reaction of enecarbamates with nucleophiles or electrophiles. For example, active methylene compounds add to the α-position of enecarbamates under the conditions using Brönstead acids as catalysts.

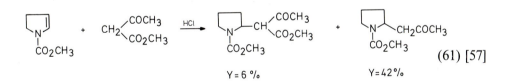

(61) [57]

The Friedel-Crafts reaction of the enecarbamates gives β-acylated enecarbamates [76].

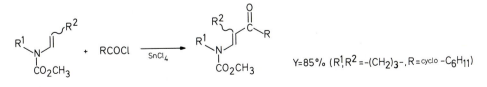

Y=85% (R^1,R^2 =-(CH$_2$)$_3$-, R=cyclo -C$_6$H$_{11}$)

(62)

Table 7. Formation of Enecarbamates and Enamides from Anodically Prepared α-Methoxycarbamates and α-Methoxyamides

Carbamate, amide	Catalyst	Product	Yield (%)	Ref.
7-membered ring lactam, N–OCH$_3$, C=O	none	7-membered ring, NH, C=O	55	[35]
pyrrolidine, N–OCH$_3$, N–CHO	NH$_4$Br	pyrroline, N–CHO	76	[75]
morpholine-type ring, O, N–OCH$_3$, N–CHO	NH$_4$Br	morpholine enamide, N–CHO	73	[75]
OHC–N⟩OCH$_3$⟨N–CHO	NH$_4$Br	OHC–N⟩=⟨N–CHO	46	[75]
$CH_3-\overset{NHCOCH_3}{\underset{CO_2C_2H_5}{C}}-OCH_3$	NH$_4$Br	$CH_2=C\overset{NHCOCH_3}{\underset{CO_2C_2H_5}{}}$	76	[56]
$C_3H_7CH\overset{C_4H_9}{\underset{OCH_3}{}}NCO_2CH_3$	NH$_4$Cl	$C_2H_5CH=CH\overset{C_4H_9}{}NCO_2CH_3$	94	[76]
pyrrolidine, N–OCH$_3$, N–CO$_2$CH$_3$	NH$_4$Cl	pyrroline, N–CO$_2$CH$_3$	91	[76]
$CH_3O–$pyrrolidine, CO$_2$CH$_3$, N–CO$_2$CH$_3$	NH$_4$Cl	pyrroline, CO$_2$CH$_3$, N–CO$_2$CH$_3$	70	[78]
$CH_3O–$piperidine, N–OCH$_3$, CO$_2$CH$_3$	NH$_4$Cl	dihydropyridine, N–CO$_2$CH$_3$	86	[78]
$CH_3O–$pyrrolidine, N–OCH$_3$, CO$_2$CH$_3$	NH$_4$Cl	pyrrole, N–CO$_2$CH$_3$	62	[78]

Similar to the Friedel-Crafts reaction, the Vilsmeyer reaction yields β-formyl-enecarbamates [76].

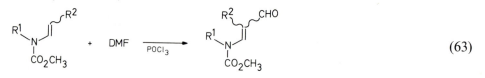

$$(63)$$

β-Formyl-enecarbamates are useful starting materials in the synthesis of heterocyclic compounds as exemplified below.

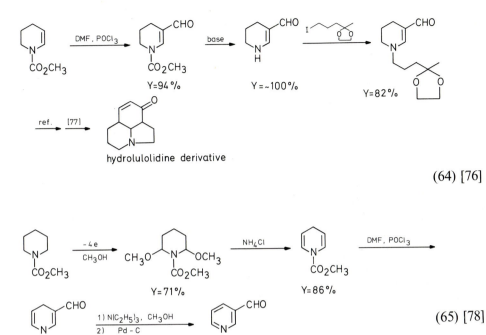

hydrolulolidine derivative

(64) [76]

(65) [78]

The hydroboration of enecarbamates gives the corresponding β-hydroxy-carbamates in reasonable yields [78].

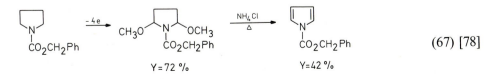

Pyrrole derivatives are obtained by dimethoxylation of pyrrolidine-carbamates followed by elimination of two molecules of methanol.

(67) [78]

References

1. Mann, C. K.: Anal. Chem. *36*, 2425 (1964)
2. a) Portis, L. C., Bhat, V. V., Mann, C. K.: J. Org. Chem. *35*, 2175 (1970)
   b) Ross, S. D.: Tetrahedron Lett. *1973*, 1237
   c) Portis, L. C., Klug, J. T., Mann, C. K.: J] Org. Chem. *39*, 3488 (1974)
3. a) Hampson, N. A., Lee, J. B., Morley, J. R., Scanlon, B.: Can. J. Chem. *47*, 3729 (1969)
   b) Hampson, N. A., Lee, J. B., Morley, J. R., McDonald, K. I., Scanlon, B.: Tetrahedron *26*, 1109 (1970)
4. Robertson, P. M., Schwager, F., Ibil, N.: Electrochim. Acta *18*, 923 (1973)
5. Shono, T., Toda, T., Oshino, N.: J. Am. Chem. Soc. *104*, 2639 (1982)
6. Shono, T., Toda, T., Oshino, N.: Drug Metab. Dispos. *9*, 481 (1981)
7. Barry, J. E., Finkelstein, M., Mayeda, E. A., Ross, S. D.: J. Org. Chem. *39*, 2695 (1974)
8. Chiba, T., Takata, Y.: J. Org. Chem. *42*, 2973 (1977)
9. Bidan, G., M.: Tetrahedron *37*, 2297 (1981)
10. a) Nelson, S. F., Hintz, P. J.: J. Am. Chem. Soc. *94*, 7108, 7114 (1972)
    See also b) Eismer, U., Zemer, Y.: J. Electroanal. Chem. *34*, 81 (1972)
    c) Karabinas, P., Heitbaum, J.: J. Electroanal. Chem. *76*, 235 (1977)
    d) Stetter, J. R., Blurton, K. F., Valentine, A. M., Tellefsen, K. A.: J. Electrochem. Soc. *125*, 1804 (1978)
    e) Kinlen, P. J., Evans, D. H.: J. Electroanal. Chem. *129*, 149 (1981)
11. Blackham, A. U., Kwak, S., Palmer, J. L.: J. Electrochem. Soc. *122*, 1081 (1975)
12. Bauer, R., Wendt, H.: Angew. Chem. *90*, 214 (1978)
13. Lauke, B. L., Asirvatham, M. R., Mann, C. K.: J. Org. Chem. *42*, 670 (1977)
14. Fuchigami, T., Iwaoka, T., Nonaka, T., Sekine, T.: Bull. Chem. Soc. Jpn. *53*, 2040 (1980)
15. Kossai, R., Simonet, J.: Tetrahedron Lett. *21*, 3575 (1980)
16. Wamhoff, H., Kunz, G.: Angew. Chem. Int. Ed. Engl. *20*, 797 (1981)
17. Masui, M., Ozaki, S.: Tetrahedron. Lett. *23*, 2867 (1982)
18. Bauer, R., Wendt, H.: Angew. Chem. Int. Ed. Engl. *17*, 370 (1978)
19. a) Hand, R. L., Nelson, R. F.: J. Am. Chem. Soc. *96*, 850 (1974)
    b) Hand, R. L., Nelson, R. F.: J. Electrochem. Soc. *125*, 1059 (1978)
20. Bacon, J., Adams, R. N.: J. Am. Chem. Soc. *90*, 6596 (1968)
21. a) Wawzonek, S., McIntyre, T. W.: J. Electrochem. Soc. *114*, 1025 (1967)
    b) Wawzonek, S., IcIntyre, T. W.: J. Electrochem. Soc. *119*, 1270 (1972)
22. a) Vettorazzi, N., Silber, J. J., Sereno, L.: J. Electronanal. Chem. *125*, 459 (1981)
    b) Desideri, G., Lepri, L., Heimler, D.: J. Electroanal. Chem. *52*, 105 (1974)
    c) Barbey, G., Delahaye, D., Caullet, C.: Bull. Soc. Chim. Fr. *1971*, 3377
    d) Yasukochi, K., Taniguchi, I., Yamaguchi, H., Tanoue, T.: Bull. Chem. Soc. Jpn. *52*, 1573 (1979)
23. Matsuda, Y., Shono, A., Iwakura, C., Ohshiro, Y., Agawa, T., Tamura, H.: Bull. Chem. Soc. Jpn. *44*, 2960 (1971)
24. Volkov, A., Tourillon, G., Lacaze, P. C., Dubois, J. E.: J. Electroanal. Chem. *115*, 279 (1980)
25. a) Leedy, D. W., Adams, R. N.: J. Am. Chem. Soc. *92*, 1646 (1970)
    b) Reynolds, R., Line, L. L., Nelson, R. F.: J. Am. Chem. Soc. *96*, 1087 (1974)
    c) Svanholm, U., Parker, V. D.: J. Am. Chem. Soc. *96*, 1234 (1974)
    d) Yoshida, K., Fueno, T.: J. Org. Chem. *41*, 731 (1976)
26. Yoshida, K., Fueno, T.: J. Org. Chem. *37*, 4145 (1972)

27. Andreades, S., Zahnow, E. W.: J. Am. Chem. Soc. *91*, 4181 (1969)
28. Shono, T., Matsumura, Y., Inoue, K., Ohmizu, H., Kashimura, S.: J. Am. Chem. Soc. *104*, 5753 (1982)
29. Mizuno, S., Nanya, S.: Bull. Nagoya Inst. Technol. *6*, 275 (1954)
30. Mizuno, S.: Bull. Nagoya Inst. Technol. *12*, 47 (1960)
31. Mizuno, S.: Bull. Nagoya Inst. Technol. *13*, 65 (1961)
32. Couch, D. E.: Electrochim. Acta *9*, 327 (1964)
33. a) Ross, S. D., Finkelstein, M., Petersen, R. D.: J. Am. Chem. Soc. *86*, 2745 (1964)
33. b) Ross, S. D., Finkelstein, M., Petersen, R. C.: J. Org. Chem. *31*, 128 (1966)
34. Ross, S. D., Finkelstein, M., Rudd, E. J.: J. Org. Chem. *37*, 2387 (1972)
35. Shono, T., Matsumura, Y., Hamaguchi, H.: The 6th Symposium on Oxidation, the Chemical Society of Japan (1972), Abstract, p. 43
36. Arita, S., Hirai, N., Nishimura, Y., Takeshita, K.: Nippon Kagaku Kaishi *1973*, 2160
37. a) Shono, T., Hamaguchi, H., Matsumura, Y.: J. Am. Chem. Soc. *97*, 4264 (1976)
    b) See also Blum, Z., Nyberg, K.: Acta Chem. Scand. *B36*, 165 (1982)
38. Genies, M.: Bull. Soc. Chim. Fr. *1975*, 390
39. a) Cedheim, L., Eberson, L., Helgée, B., Nyberg, K., Servin, R., Sternerup, H.: Acta Chem. Scand. *B29*, 617 (1975)
    b) See also Cedheim, L., Eberson, L., Helgée, B., Nyberg, K., Servin, R., Sternerup, H.: Acta Chem. Scand. *B33*, 113 (1979)
40. a) Nyberg, K., Servin, R.: Acta Chem. Scand. *B30*, 640 (1976)
    b) See also Finkelstein, M., Nyberg, K., Ross, S. D., Servin, R.: Acta Chem. Scand. *B32*, 182 (1978)
41. Mitzlaff, M., Warning, K., Jensen, H.: Justus Liebigs Ann. Chem. *1978*, 1713
42. Torii, S., Yamanaka, T., Tanaka, H.: J. Org. Chem. *43*, 2882 (1978)
43. Okita, M., Wakamatsu, T., Ban, Y.: J. Chem. Soc., Chem. Commun. *1979*, 749
44. Irie, K., Okita, M., Wakamatsu, T., Ban, Y.: Bouveau J. Chimie *4*, 275 (1980)
45. Shono, T., Matsumura, Y., Tsuda, K., Uchida, K.: The 43rd Annual Meeting of the Chemical Society of Japan (1981), Abstract, p. 1086
46. Ross, S. D., Finkelstein, M., Petersen, R. C.: J. Am. Chem. Soc. *88*, 4657 (1966)
47. Finkelstein, M., Ross, S. D.: Tetrahedron *28*, 4497 (1972)
48. Rudd, E. J., Finkelstein, M., Ross, S. D.: J. Org. Chem. *37*, 1763 (1972)
49. Shono, T., Matsumura, Y., Uchida, K., Tsuda, K.: unpublished data.
50. a) Hellmann, H.: Angew. Chem. *69*, 463 (1957)
    b) Zaugg, H. E., Martin, W. B.: Org. React. *14*, 52 (1965)
    c) Zaugg, H. E.: Synthesis *1970*, 49
    d) Mathieu, J., Weill-Raynal, J.: Formation of C—C Bonds, Vol. III, p. 248, Stuttgart, Georg Thieme 1977
51. Ross, S. D., Finkelstein, M., Petersen, R. C.: J. Org. Chem. *31*, 133 (1966)
52. Amidoalkylation using the starting materials prepared by the usual chemical methods are not described in this book.
53. Nyberg, K.: Acta Chem. Scand. *B28*, 825 (1974)
54. Malmberg, M., Nyberg, K.: Acta Chem. Scand. *B33*, 69 (1979)
55. Malmberg, M., Nyberg, K.: Acta Chem. Scand. *B35*, 411 (1981)
56. Iwasaki, T., Horikawa, H., Matsumoto, K.: Bull. Soc. Chem. Jpn. *52*, 826 (1979)
57. Shono, T., Matsumura, Y., Tsubata, K.: J. Am. Chem. Soc. *103*, 1172 (1981)
58. Shono, T., Matsumura, Y., Tsubata, K., Takata, J.: Chem. Lett. *1981*, 1121
59. Asher, V., Becu, C., Anteunis, Mar J. O., Callens, R.: Tetrahedron Lett. *22*, 141 (1981)

60. Shono, T., Matsumura, Y., Tsubata, K., Inoue, K., Nishida, R.: Chem. Lett. *1983*, 21
61. Shono, T., Matsumura, Y., Tsubata, K., Inoue, K., Onomura, O.: Proceeding of the 42nd Symposium on the Synthetic Organic Chemistry (1982), Abstract, p. 13
62. Malmberg, M., Nyberg, K.: J. Chem. Soc., Chem. Commun. *1979*, 167
63. Shono, T., Matsumura, Y., Tsubata, K.: Tetrahedron Lett. *22*, 2411 (1981)
64. Shono, T., Matsumura, Y., Tsubata, K.: The 43rd Annual Meeting of the Chemical Society of Japan (1981), Abstract, p. 1085
65. Shono, T., Matsumura, Y., Tsubata, K., Uchida, K., Kobayashi, H.: The 47th Annual Meeting of the Chemical Society of Japan (1983), Abstract, p. 701
66. Shono, T., Matsumura, Y., Tsubata, K.: Tetrahedron Lett. *22*, 3249 (1981)
67. Shono, T., Matsumura, Y., Kanazawa, T.: The 45th Annual Meeting of the Chemical Society of Japan (1982), Abstract, p. 1028
68. Warning, K., Mitzlaff, M.: Tetrahedron Lett. *1979*, 1565
69. a) Nishitani, T., Horikawa, H., Iawasaki, T., Matsumoto, K., Inoue, I., Miyoshi, M.: J. Org. Chem. *47*, 1706 (1982)
    b) See also Nishitani, T., Iawasaki, T., Mushika, Y., Inoue, I., Miyoshi, M.: Chem. Pharm. Bull. *28*, 1137 (1980)
70. Warning, K., Mitzlaff, M.: Tetrahedon Lett. *1979*, 1563
71. Shono, T., Matsumura, Y., Kashimura, S.: J. Org. Chem. *48*, 3338 (1983)
72. Warning, K., Mitzlaff, M., Jensen, H.: Justus Liebigs Ann. Chem. *1978*, 1707
73. Shono, T., Matsumura, Y., Kanazawa, T.: Tetrahedron Lett. *24*, 1259 (1983)
74. Shono, T., Matsumura, Y., Kanazawa, T., Habuka, M., Toyoda, K.: The 47th Annual Meeting of the Chemical Society of Japan (1983), Abstract, p. 937
75. Nyberg, K.: Synthesis *1976*, 545
76. Shono, T., Matsumura, Y., Tsubata, K., Sugihara, Y.: Tetrahedron Lett. *23*, 1201 (1982)
77. Wenkert, E., Dave, K. G., Stevens, R. V.: J. Am. Chem. Soc. *90*, 6177 (1968)
78. Shono, T., Matsumura, Y., Tsubata, K., Sugihara, Y., Yamane, S., Kanazawa, T., Aoki, T.: J. Am. Chem. Soc. *104*, 6697 (1982)

2.7 Oxidation of Carbanions and Carboxylate Anions

2.7.1 Oxidation of Carbanions

As the oxidation potential clearly shows, carbanions may easily be oxidized by the anodic method (Table 1).

The most typical process of the anodic oxidation of carbanions is the formation of radical species.

$$R^- \xrightarrow{\ -e\ } R\cdot \qquad\qquad (1)$$

Although generally dimerization is one of the typical reactions of radical species, the yields of the dimers are not always high from the viewpoint of organic synthesis. Thus the anodic oxidation of anions of monoalkylated malonic esters in acetonitrile yields the corresponding dimers in 20–55% yield [4].

Table 1. Oxidation Potential of Carbanions

Compound	Solvent	Potential	Ref.	Compound	Solvent	Potential	Ref.
$(CH_3)_2Mg$	1,2-Dimethoxy-ethane	E 1/2, -1.2 (Ag/Ag$^+$)	[1]	$CH_3CH{<}^{CHO}_{CN}$	DMSO	E 1/2, -0.06 (Fc/Fc$^+$)^a	[3]
$(C_6H_5CH_2)_2Mg$	1,2-Dimethoxy-ethane	E 1/2, -1.2 (Ag/Ag$^+$)	[1]	(cyclohexanone)—CN	DMSO	E 1/2, -0.23 (Fc/Fc$^+$)^a	[3]
$(C_6H_5)_3CLi$	THF	E$_p$, -1.33 (SCE)	[2]	$C_6H_5COCH_2CN$	DMSO	E 1/2, 0.06 (Fc/Fc$^+$)^a	[3]
$(C_6H_5)_2CHLi$	THF	E$_p$, -1.37 (SCE)	[2]	$CH_3COCH_2COCH_3$	DMSO	E 1/2, 0.49 (Fc/Fc$^+$)^a	[3]
$C_6H_5CH_2Li$	THF	E$_p$, -1.45 (SCE)	[2]	$CH_3COCH_2CO_2C_2H_5$	DMSO	E 1/2, 0.38 (Fc/Fc$^+$)^a	[3]
$CH_2{=}CHCH_2Li$	THF	E$_p$, -1.40 (SCE)	[2]	$CH_2(CO_2C_2H_5)_2$	DMSO	E 1/2, 0.39 (Fc/Fc$^+$)^a	[3]

^{a}Fc: Ferrocene

$$2 C_2H_5CH(CO_2C_2H_5)_2 \xrightarrow[C_2H_5ONa, CH_3CN]{-2e} (C_2H_5O_2C)_2 \overset{C_2H_5}{\underset{C_2H_5}{C-C}}(CO_2C_2H_5)_2 \qquad (2)$$
$$Y=55\%$$

Acetylide anions are dimerized to diacetylenes through anodic oxidation [5].

$$2 \; C_6H_5{-}C{\equiv}CLi \xrightarrow[THF]{-2e} C_6H_5{-}C{\equiv}C{-}C{\equiv}C{-}C_6H_5 \qquad (3)$$
$$Y=35\%$$

Grignard reagents give rather satisfactory results in anodic dimerization [6].

$$2 RMgBr \xrightarrow{-2e} R{-}R$$

$R = C_5H_{11}$ Y=55-60 %
$R = C_{18}H_{37}$ Y=54 %
$R = C_6H_5$ Y=55 %

$$(4)$$

The anodically generated radical species may be trapped by olefins to give corresponding adducts (Table 2)

81

Table 2. Formation of Adducts from Anodically Generated Radicals and Olefins

Anion	Olefin	Product	Yield (%)	Ref.
$CH_3CO\bar{C}HCO\,CH_3$	$CH_2{=}CHOC_2H_5$		36	[7]
$\bar{C}H(CO_2CH_3)_2$	$CH_2{=}CHOC_2H_5$	$(CH_3O_2C)_2CHCH_2CH{\begin{smallmatrix}OC_2H_5\\OCH_3\end{smallmatrix}}$	37	[7]
$(CH_3)_2\bar{C}NO_2$	$CH_2{=}CH{-}\bigcirc$	$(CH_3)_2\overset{NO_2}{\underset{}{C}}{-}CH_2CH{\begin{smallmatrix}\bigcirc\\OCH_3\end{smallmatrix}}$	43	[8]
C_4H_9MgBr	$CH_2{=}CH{-}\bigcirc$	$(C_4H_9CH_2\underset{I}{C}H{-}\bigcirc)_2$	29	[9]

2.7.2 Oxidation of Carboxylate Anions

Oxidation of carboxylic acids can be classified into two major categories, formation of radical intermediates followed by dimerization and generation of cation intermediates followed by reaction with nucleophiles.

$$RCO_2^- \xrightarrow{-e} RCO_2\cdot \xrightarrow{-CO_2} R\cdot \begin{cases} \longrightarrow R{-}R \\ \xrightarrow{-e} R^+ \xrightarrow{Nu^-} R{-}Nu \end{cases} \tag{5}$$

The reaction is controlled by a variety of factors including anode material, anode potential, current density, solvent, supporting electrolyte, structure of R, and temperature.

2.7.2.1 Formation of Radicals. Kolbe-Type Reactions

The Kolbe dimerization is believed to be favored by the following reaction conditions: high concentration of carboxylic acid, low pH value, absence of foreign anions, high current density, and use of a platinum anode. The Kolbe type reaction has been reviewed in detail [10, 11]. In the Kolbe dimerization, retention of the configuration of the alkyl radical R· has never been observed. For example, in the following two reactions, the obtained products are completely racemic.

$$\underset{C_2H_5}{\overset{CH_3}{>}}\overset{*}{C}\underset{CO_2^-}{\overset{CO_2C_2H_5}{<}} + \underset{CH_3}{\overset{CH_3}{>}}CHCH_2CO_2^- \xrightarrow[CH_3OH,\ KOH]{-2e} \underset{C_2H_5}{\overset{CH_3}{>}}C\underset{CH_2CH<}{\overset{CO_2C_2H_5}{<}}\begin{smallmatrix}CH_3\\CH_3\end{smallmatrix} \qquad (6)\ [12]$$

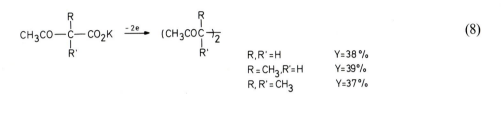

$$\begin{matrix} CH_3 \\ C_2H_5 \end{matrix} \overset{*}{C}HCO_2^- \quad + \quad CH_3O\overset{O}{\overset{\|}{C}}CH_2CO_2^- \quad \xrightarrow[CH_3OH,\ KOH]{-2e} \quad \begin{matrix} CH_3 \\ C_2H_5 \end{matrix} CHCH_2CO_2CH_3 \qquad (7)\ [13]$$

The course of the reaction is greatly affected by the solvent used:
In water-dioxane (5:3)

$$CH_3CO-\underset{R'}{\overset{R}{\underset{|}{C}}}-CO_2K \quad \xrightarrow{-2e} \quad (CH_3CO\underset{R'}{\overset{R}{\underset{|}{C}}})_2 \qquad (8)$$

R,R' =H	Y=38%
R = CH_3,R'=H	Y=39%
R, R' = CH_3	Y=37%

$$CH_3CO-\underset{CH_3}{\overset{CH_3}{\underset{|}{\overset{|}{C}}}}-CO_2K \quad \xrightarrow{-2e} \quad CH_3COC\overset{CH_2}{\underset{CH_3}{\diagdown}} \qquad Y=90\% \qquad (9)$$

In acetonitrile

$$CH_3CO-\underset{CH_3}{\overset{CH_3}{\underset{|}{\overset{|}{C}}}}-CO_2K \quad \xrightarrow{-2e} \quad CH_3CO\underset{CH_3}{\overset{CH_3}{\underset{|}{\overset{|}{C}}}}NHCOCH_3 \qquad Y=50\% \qquad (10)$$

The presence of a foreign anion may disturb the formation of the dimer [15].
   Since the Kolbe dimerization has already been reviewed [16], only few examples of its application to the synthesis of pheromones are given in the following.

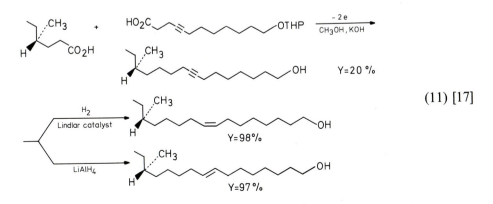

$$(11)\ [17]$$

2. Anodic Oxidations

$$\underset{\text{CH}_3}{\overset{\text{CH}_3}{\text{C}_{18}\text{H}_{37}\text{CH}(\text{CH}_2)_4\text{CO}_2\text{H}}} \ + \ \underset{}{\overset{\text{CH}_3}{\text{CH}_3\text{COCH}(\text{CH}_2)_3\text{CO}_2\text{H}}} \ \xrightarrow[\text{CH}_3\text{OH, KOH}]{-2e} \ \underset{}{\overset{\text{CH}_3 \quad \text{CH}_3}{\text{C}_{18}\text{H}_{37}\text{CH}(\text{CH}_2)_7\text{CHCOCH}_3}}$$

Y=42.5%

$$\text{C}_4\text{H}_9\text{C}\!\equiv\!\text{C}(\text{CH}_2)_3\text{CO}_2\text{H} \ + \ \text{HO}_2\text{C}(\text{CH}_2)_6\text{CO}_2\text{CH}_3 \ \xrightarrow[\text{CH}_3\text{OH, KOH}]{-2e} \ \text{C}_4\text{H}_9\text{C}\!\equiv\!\text{C}(\text{CH}_2)_9\text{CO}_2\text{CH}_3$$

Y=59%

(12) [18]

$$\xrightarrow{\begin{array}{l}\text{1) }\text{H}_2/\text{Pd}\\ \text{2) LiAlH}_4\\ \text{3) }\text{CH}_3\text{COCl}\end{array}} \ \text{C}_4\text{H}_9\text{CH}\!=\!\text{CH}(\text{CH}_2)_{10}\text{OCOCH}_3 \qquad Z-\text{isomer}$$

(13) [19]

The radical species generated by the anodic oxidation of carboxylic acids can be trapped by oxygen [20] or active olefins.

$$2\,\text{CF}_3\text{CO}_2\text{H} \ + \ 2\,\text{CH}_2\!=\!\text{CHCOCH}_3 \ \xrightarrow[\text{CH}_3\text{CN, H}_2\text{O, NaOH}]{-2e}$$

Y=24%

(14) [21]

$$2\,\text{CH}_3\text{CO}_2\text{H} \ + \ \text{C}_2\text{H}_5\text{O}_2\text{CCH}\!=\!\text{CHCO}_2\text{C}_2\text{H}_5 \ \xrightarrow[\text{CH}_3\text{CN, H}_2\text{O, NaOH}]{-2e}$$

Y=80%

(15) [22]

2.7.2.2 Formation of Cations

When the cation R^+ is adequately stable and the reaction conditions are favorable for its formation, the radical $R\cdot$ formed from carboxylic acid RCO_2H is further oxidized to the cation R^+ which is then trapped by a nucleophile, Nu^- [11, 23]. This reaction is applied to the transformation of a carboxy group to a hydroxy group.

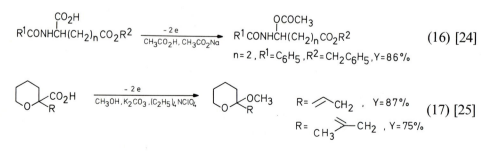

$$\underset{}{\overset{\text{CO}_2\text{H}}{\text{R}^1\text{CONHCH}(\text{CH}_2)_n\text{CO}_2\text{R}^2}} \ \xrightarrow[\text{CH}_3\text{CO}_2\text{H, CH}_3\text{CO}_2\text{Na}]{-2e} \ \underset{}{\overset{\text{OCOCH}_3}{\text{R}^1\text{CONHCH}(\text{CH}_2)_n\text{CO}_2\text{R}^2}}$$

(16) [24]

n= 2, R^1= C_6H_5, R^2= $CH_2C_6H_5$, Y=86%

$$\xrightarrow[\text{CH}_3\text{OH, K}_2\text{CO}_3, (\text{C}_2\text{H}_5)_4\text{NClO}_4]{-2e}$$

R= —CH₂, Y=87%

R= —CH₂, Y=75%

(17) [25]

The formation of cations is utilized for the synthesis of unsaturated compounds.

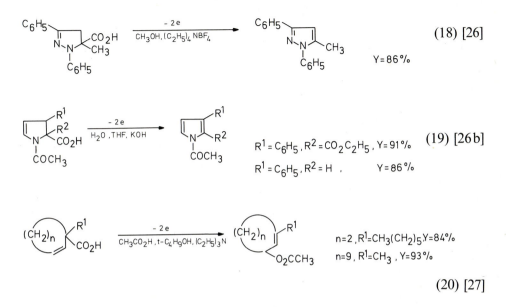

$$\text{(18) [26]} \quad Y=86\%$$

$$R^1 = C_6H_5, R^2 = CO_2C_2H_5, Y=91\% \quad \text{(19) [26b]}$$
$$R^1 = C_6H_5, R^2 = H, \quad Y=86\%$$

$$n=2, R^1=CH_3(CH_2)_5 Y=84\%$$
$$n=9, R^1=CH_3, Y=93\%$$

$$\text{(20) [27]}$$

A Wagner-Meerwein type rearrangement of the cation has been often observed in the oxidation of carboxylic acids.

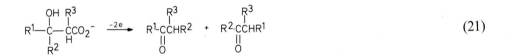

$$\text{(21)}$$

The relative migratory aptitude of R^1 and R^2 has been studied, and this type of rearrangement has been applied to the synthesis of *dl*-muscone [28].

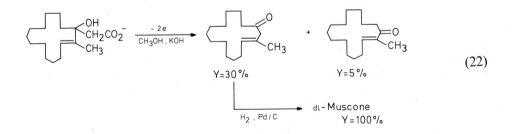

$$Y=30\% \qquad Y=5\%$$

$$\xrightarrow[H_2, Pd/C]{} \quad dl\text{-Muscone} \quad Y=100\%$$

$$\text{(22)}$$

The intermediate cation may be oxidized to the corresponding carbonyl compound.

2. Anodic Oxidations

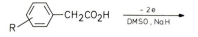

$$R\text{—}\langle\rangle\text{—}CH_2CO_2H \xrightarrow[\text{DMSO, NaH}]{-2e} R\text{—}\langle\rangle\text{—}CHO \qquad (23)\ [29]$$

R=4-CH₃O : Y=72%, R=4-Cl : Y=62%
R=3-CH₃O : Y=78%, R=H : Y=41%

References

1. Psarras, T., Dessy, R. E.: J. Am. Chem. Soc. *88*, 5132 (1966)
2. Jaun, B., Schwarz, J., Breslow, R.: J. Am. Chem. Soc. *102*, 5741 (1980)
3. a) Kern, J. M., Federlin, P.: Tetrahedron Lett. *1977*, 837
   See also b) Kern, J. M., Federlin, P.: J. Electroanal. Chem. *96*, 209 (1979)
   c) Lochert, P., Federlin, P.: Tetrahedron Lett. *1973*, 1109
   d) Van den Born, H. W., Evans, D. H.: J. Am. Chem. Soc. *96*, 4296 (1974)
4. a) Thomas, H. G., Streukens, M., Peek, R.: Tetrahedron Lett. *1978*, 45
   See also b) Binns, T. D., Brettle, R.: J. Chem. Soc. (C) *1966*, 336
   c) Brettle, R., Parkin, J. G.: J. Chem. Soc. (C) *1967*, 1352
   d) Brettle, R., Seddon, D.: J. Chem. Soc. (C) *1970*, 1153
   e) Brettle, R., Parkin, J. G., Seddon, D.: J. Chem. Soc. (C) *1970*, 1317
   f) Brettle, R., Seddon, D.: J. Chem. Soc. (C) *1970*, 2175
5. Bauer, R., Wendt, H.: J. Electroanal. Chem. *80*, 395 (1977)
6. Morgat, J. L., Pallaud, R.: C.R. Acad. Sci. *260*, 574, 5579 (1965)
7. Schäfer, H. J., Azrak, A. Al.: Chem. Ber. *105*, 2398 (1972)
8. Schäfer, H. J.: Angew. Chem. Int. Ed. Engl. *20*, 911 (1981)
9. Schäfer, H. J., Küntzel, H.: Tetrahedron Lett. *1970*, 3333
10. Eberson, L.: Chemistry of Carboxylic Acids and Esters (ed.) Patai, S., p. 53, London, John Wiley & Sons 1969
11. Utley, J. H. P.: Technique of Electroorganic Synthesis (ed.) Weinberg, N. L., p. 793, Chapter VI, New York, John Wiley & Sons 1974
12. Eberson, L., Petterson, G. R.: Acta Chem. Scand. *B27*, 1159 (1973)
13. Eberson, L., Nyberg, K., Servin, S., Wennerbeck, L.: Acta Chem. Scand. *B30*, 186 (1976)
14. Chkir, M., Lelandols, D., Bacquet, C.: Can. J. Chem. *59*, 945 (1981)
15. Coleman, J. P., Utley, J. H. P., Weedon, B. C. L.: J. Chem. Soc., Chem. Commun. *1971*, 438
16. For example, Schäfer, H. J.: Angew. Chem. Int. Ed. Engl. *20*, 911 (1981)
17. Jensen, U., Schäfer, H. J.: Chem. Ber. *114*, 292 (1981)
18. Seidel, W., Schäfer, H. J.: Chem. Ber. *113*, 451 (1980)
19. Seidel, W., Schäfer, H. J.: Chem. Ber. *113*, 3898 (1980)
20. Barry, J. E., Finkelstein, M., Mayeda, E. A., Ross, S. D.: J. Am. Chem. Soc. *98*, 8098 (1976)
21. Renaud, R. N., Champagne, P. J., Savard, M.: Can. J. Chem. *57*, 2617 (1979)
22. a) Renaud, R. N., Champagne, P. J.: Can. J. Chem. *57*, 990 (1979)
    b) Renaus, R. N., Champagne, P. J.: Can. J. Chem. *58*, 1101 (1980)
23. Wawzonek, S.: Synthesis *1971*, 285
24. a) Nishitani, T., Iwasaki, T., Mushika, Y., Inoue, I., Miyoshi, M.: Chem. Pharm. Bull. *28*, 1137 (1980)
    b) Iwasaki, T., Horikawa, H., Matsumoto, K., Miyoshi, M.: J. Org. Chem., *42*, 2419 (1977)

25. Wuts, P. G. M., Sutherland, C.: Tetrahedron Lett. *23*, 3987 (1982)
26. a) Diaz, A.: J. Org. Chem. *42*, 3949 (1977)
    b) See also Horikawa, H., Iwasaki, T., Matsumoto, K., Miyoshi, M.: J. Org. Chem. *43*, 335 (1978)
27. Torii, S., Inokuchi, T., Mizukuchi, K., Yamazaki, M.: J. Org. Chem. *44*, 2303 (1979)
28. Shono, T., Hayashi, J., Omoto, H., Matsumura, Y.: Tetrahedron Lett. *1977*, 2667
29. Mandell, L., Daley, R. F., Day, R. A., Jr.: J. Org. Chem. *42*, 1461 (1977)

2.8 Oxidation of Aromatic Systems

Removal of electrons from aromatic π-electron systems may be achieved by electrochemical oxidation, and the resulting aromatic cation radical or other aromatic cationic species undergo interesting and important reactions. Typical reactions are aromatic substitution (1) and coupling (2).

$$ArH \xrightarrow[Nu^-]{-2e} Ar-Nu + H^+ \tag{1}$$

$$2\,ArH \xrightarrow{-2e} Ar-Ar + 2\,H^+ \tag{2}$$

2.8.1 Aromatic Substitution

2.8.1.1 Acetoxylation

Since the discovery of the anodic aromatic nuclear acetoxylation in the study on the Kolbe electrolysis carried out in the presence of naphthalene [1], it has been studied extensively from both the mechanistic and synthetic point of view [2]. Although a variety of mechanisms including radical substitution and EE mechanisms have been proposed, an ECEC mechanism (3) is now believed to be most probable.

$$ArH \xrightarrow{-e} ArH^{+\cdot} \xrightarrow{Nu^-} Ar\!\!\begin{smallmatrix}\cdot\,/H\\\backslash Nu\end{smallmatrix} \xrightarrow{-e} Ar\!\!\begin{smallmatrix}+\,/H\\\backslash Nu\end{smallmatrix} \xrightarrow{-H^+} ArNu \tag{3}$$

When the aromatic substrates contain benzylic hydrogens, anodic benzylic substitution always competes with nuclear substitution. Some typical examples of nuclear acetoxylation are shown in Table 1. However, only low yields of acetoxylated products are generally obtained.

As Table 2 reveals the yields of nuclear hydroxylation products are improved by carrying out the reaction in the system of CF_3CO_2Na, CF_3CO_2H, since the first products, namely trifluoroacetoxylated compounds, are generally stable under the conditions of anodic nuclear substitution.

2. Anodic Oxidations

Table 1. Nuclear Acetoxylation

Substrate	Solvent and supporting electrolyte	Product	Yield (%)	Ref.
(naphthalene)	$CH_3CO_2H-CH_3CO_2Na$	(naphthalene)$-O_2CCH_3$	24	[3] [4]
(benzene)$-OCH_3$	$CH_3CO_2H-CH_3CO_2Na$	CH_3O-(benzene)$-O_2CCH_3$	27 ($o:p=9:1$)	[4]
(biphenyl-CO_2H)	$CH_3CO_2H-CH_3CO_2Na$	(dibenzo lactone)	36	[4]
H_3C-(benzene with CH_3, CH_3)	$CH_3CO_2H-CH_3CO_2K$	CH_3-(benzene with CH_3, CH_3)$-O_2CCH_3$	12.6	[5]
CH_3O-(benzene)$-OCH_3$	$CH_3CO_2H-CH_3CO_2Na$	CH_3O-(benzene)$-OCH_3$, O_2CCH_3	68	[6]
$F-$(benzene)$-OCH_3$	$CH_3CO_2H-CH_3CO_2K$	CH_3CO_2-(benzene)$-OCH_3$	70-80	[7]

Table 2. Nuclear Hydroxylation

Substrate	Solvent and supporting electrolyte	Product	Yield (%)	Ref.
(benzene)	$CF_3CO_2H-CF_3CO_2Na-(CF_3CO)_2O$	(benzene)$-OH$	65	[8]
(dichlorobenzene Cl, Cl)	$CF_3CO_2H-CF_3CO_2Na-(CF_3CO)_2O$	$HO-$(benzene Cl, Cl)	70	[8]
(benzene)$-CO_2C_2H_5$	$CH_2Cl_2-CF_3CO_2H-(C_2H_5)_4NBF_4$	(benzene OH)$-CO_2C_2H_5$	o, 43; p, 19	[9]
(benzene)$-CF_3$	$CF_3CO_2H-CF_3CO_2Na-(CF_3CO)_2O$	(benzene O_2CCF_3)$-CF_3$	56 ($o:p=3:10$)	[10]
(benzene)$-Cl$	$CF_3CO_2H-CF_3CO_2Na$	(benzene O_2CCF_3)$-Cl$	94 ($o:m:p=41:3:56$)	[11]
(benzene)$-NO_2$	$CF_3CO_2H-CF_3CO_2Na$	(benzene O_2CCF_3)$-NO_2$	60 ($o:m:p=22:59:19$)	[12]

When nuclear acetoxylation is performed in the presence of Pd on charcoal, the product acetoxylated at the benzylic position is reduced to the starting material; hence, the yield of nuclear acetoxylation is greatly increased, though the current yield is still low [13].

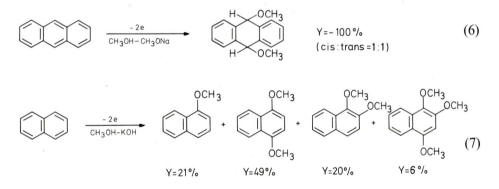

CH₃—⟨benzene⟩—CH₃ $\xrightarrow[\text{CH}_3\text{CO}_2\text{H}-\text{CH}_3\text{CO}_2\text{K , Pd/C}]{-2e}$ CH₃—⟨benzene⟩—CH₃, O₂CCH₃   Y = 78%   (4)

$\xrightarrow[\text{CH}_3\text{CO}_2\text{H}-\text{CH}_3\text{CO}_2\text{K , Pd/C}]{-2e}$ Y = 79%   (5)

2.8.1.2 Methoxylation

Nuclear methoxylation of substrates possessing rather high oxidation potentials is only achieved with difficulty while naphthalene [14] and anthracene [15] are readily methoxylated.

$\xrightarrow[\text{CH}_3\text{OH}-\text{CH}_3\text{ONa}]{-2e}$ Y = ~100% (cis : trans =1:1)   (6)

$\xrightarrow[\text{CH}_3\text{OH}-\text{KOH}]{-2e}$ Y=21% + Y=49% + Y=20% + Y=6%   (7)

The transformation of easily oxidizable substrates such as 1,4-dimethoxy-benzene to the corresponding quinone bisacetals occurs in high yields as shown in Table 3. The quinone bisacetals have been used as starting materials in a variety of organic syntheses.

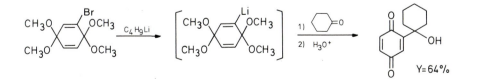

$\xrightarrow{\text{C}_4\text{H}_9\text{Li}}$ 1)   2) H₃O⁺   Y= 64%

(8) [19]

2. Anodic Oxidations

Table 3. Preparation of Quinone Bisacetals

Substrate	Solvent and supporting electrolyte	Product	Yield (%)	Ref.
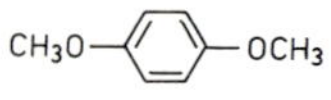 CH₃O—⟨benzene⟩—OCH₃	CH₃OH, KOH	 CH₃O,OCH₃ / CH₃O,OCH₃ (para bisacetal)	88	[16] [17]
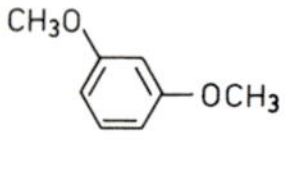 CH₃O / ⟨benzene⟩—OCH₃ (meta)	CH₃OH, KOH	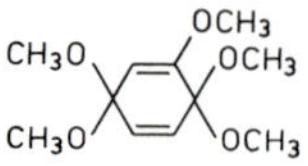 OCH₃ / CH₃O,OCH₃ / CH₃O,OCH₃	67	[16] [17]
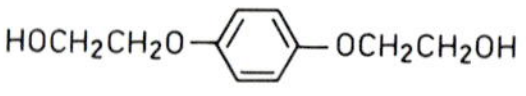 HOCH₂CH₂O—⟨benzene⟩—OCH₂CH₂OH	CH₃OH, KOH	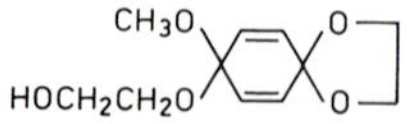 CH₃O / HOCH₂CH₂O (cyclic acetal)	~100	[18]
CH₃O—⟨benzene, Br⟩—OCH₃	CH₃OH, CH₃ONa	Br / CH₃O,OCH₃ / CH₃O,OCH₃	75	[19]
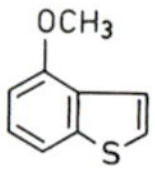 OCH₃ ⟨benzothiophene⟩	CH₃OH, KOH	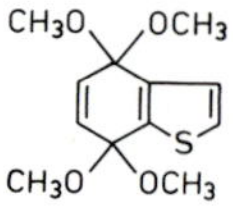 CH₃O,OCH₃ / CH₃O,OCH₃ ⟨thiophene⟩	78	[20]
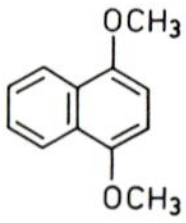 OCH₃ ⟨naphthalene⟩ OCH₃	CH₃OH, KOH	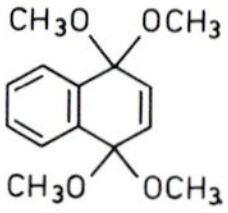 CH₃O,OCH₃ / CH₃O,OCH₃ ⟨naphthalene⟩	74	[21] [22]

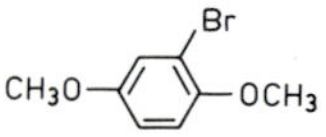

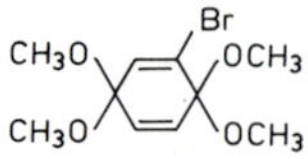

The reaction illustrated by scheme (8) has been applied to the synthesis of (±)-7,9-deoxydaunomycinone (9) [23].

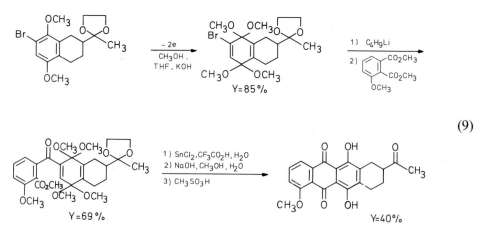

(9)

90

Further examples of using quinone bisacetals in organic synthesis are shown below.

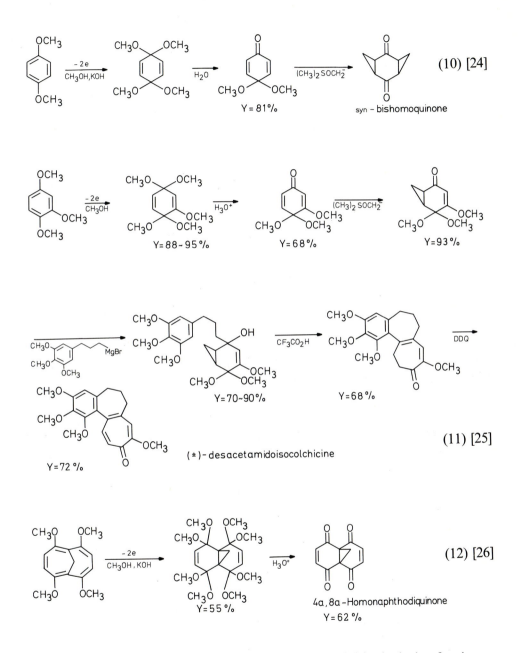

Quinone monoacetal which is obtained by careful hydrolysis of quinone bisacetal can also be prepared directly by anodic oxidation under modified reaction conditions.

2. Anodic Oxidations

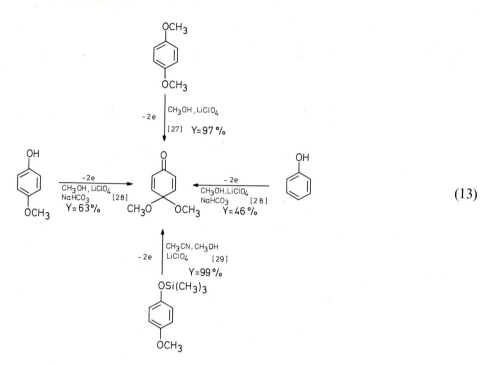

(13)

The hydrolysis of quinone bisacetals to monoacetals takes place predominantly at the less substituted side [30].

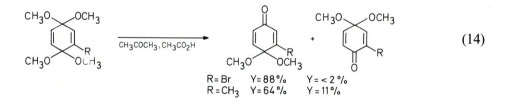

(14)

R= Br   Y= 88%   Y = < 2%
R = CH₃   Y= 64%   Y = 11%

The bromoacetal reacts with some nucleophiles.

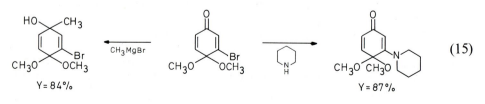

(15)

92

Quinone bisacetal also reacts with nucleophiles [31].

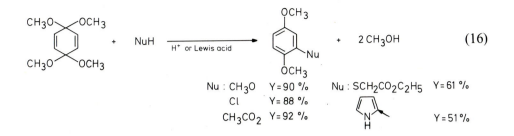

$$Nu : CH_3O \quad Y = 90\,\% \qquad Nu : SCH_2CO_2C_2H_5 \quad Y = 61\,\%$$
$$Cl \qquad Y = 88\,\%$$
$$CH_3CO_2 \quad Y = 92\,\% \qquad \qquad \qquad Y = 51\,\%$$

(16)

A new method of preparing monoacetals bearing a substituent at the side of the carbonyl group is described in scheme (17) [32].

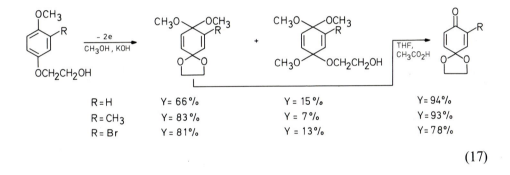

R = H	Y = 66%	Y = 15%	Y = 94%
R = CH$_3$	Y = 83%	Y = 7%	Y = 93%
R = Br	Y = 81%	Y = 13%	Y = 78%

(17)

2.8.1.3 Formation of Quinones

Besides through hydrolysis of quinone bisacetals, quinones can be prepared directly by anodic oxidation of aromatic compounds. Some examples are summarized in Table 4.

Acyl cations generated in the electrooxidative conversion of acylated hydroquinones into quinone have been utilized for the formation of esters and the Friedel-Crafts type reactions.

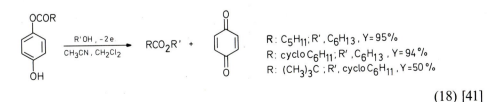

R : C_5H_{11}; R', C_6H_{13}, Y = 95%
R : cyclo C_6H_{11}; R' ,C_6H_{13} , Y = 94%
R : $(CH_3)_3C$; R', cyclo C_6H_{11} , Y = 50%

(18) [41]

Table 4. Formation of Quinones

Substrate	Solvent and supporting electrolyte	Product	Yield (%)	Ref.
phenol –OH	H_2O, H_2SO_4	benzoquinone (O=◯=O)	60	[33] [34]
phenol –OH	CH_3OH, $LiClO_4$, $NaHCO_3$	benzoquinone (O=◯=O)	37	[28]
naphthalene	DMF, H_2SO_4	naphthoquinone	20	[35]
Cl, Cl, CN, CN, OH, OH substituted benzene	CH_3CN, $LiClO_4$	Cl, Cl, CN, CN quinone	77	[36]
benzene –F_6	CF_3CO_2H, CF_3CO_2K, $(CF_3CO)_2O$	tetrafluoroquinone	75	[37]
anthracene with X (X = Br, Cl)	CH_3CN, H_2O, $LiClO_4$	anthraquinone	~100	[38]
NH_2 substituted benzene	CH_3CN, H_2O, $NaClO_4$	NH quinone imine	80–90	[39]
$OCOCH_2C_6H_5$ / OC_2H_5 naphthalene	H_2O, CH_3COCH_3	naphthoquinone	86	[40]

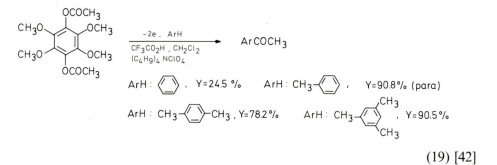

$$\text{substrate} \xrightarrow[\substack{CF_3CO_2H,\ CH_2Cl_2 \\ (C_4H_9)_4 NClO_4}]{-2e,\ ArH} ArCOCH_3$$

ArH: ◯ , Y=24.5 %   ArH: CH_3–◯ , Y=90.8 % (para)

ArH: CH_3–◯–CH_3, Y=78.2 %   ArH: CH_3–◯ with CH_3, CH_3 , Y=90.5 %

(19) [42]

2.8.1.4 Miscellaneous Oxidations

Some other oxidative reactions are briefly surveyed in Table 5.

Table 5. Miscellaneous Oxidation

Substrate	Solvent and supporting electrolyte	Product	Yield (%)	Ref.
CH_3–⬡–OH	H_2O, H_2SO_4	O=⬡(CH_3)(OH)	40	[43]
CH_3–⬡–OH	CH_3CN, CH_3OH, $(C_2H_5)_4NOH$	O=⬡(OCH_3)(CH_3)	65	[44]
C_2H_5–⬡–OH	CH_3CN, H_2O, $NaClO_4$	O=⬡(C_2H_5)(OH)	93	[45]
CH_3CO, CH_3 N–⬡–OCH_3	CH_3OH, $NaClO_4$	CH_3O–⬡ ring ClO_4^-	70	[46]

2.8.1.5 Oxidation of Benzene

Owing to the high oxidation potential of benzene, the direct anodic transformation of benzene to phenol, hydroquinone and their derivatives is not always successful. Hence, some devices are essential to achieve conversion of benzene into phenolic compounds. The anodic oxidation of benzene in trifluoroacetic acid containing sodium trifluoroacetate and subsequent hydrolysis of the reaction product affords phenol in 65% yield [8, 47].

$$\text{(20)}$$

The moderate stability of the first product, phenyl trifluoroacetate, under the conditions of anodic oxidation seems to be favorable for the preparation of phenol.

The hydroxylation of the aromatic nucleus by hydroxyl radicals generated by decomposition of hydrogen peroxide in the presence of iron(II) ions may be applied to the electrochemical synthesis of phenol from benzene since the concentration of the iron(II) ions can be controlled by the cathodic reduction of iron(III) ions formed by oxidation of iron(II) ions

2. Anodic Oxidations

with H_2O_2. Thus, in this electroreductive hydroxylation, the reaction proceeds smoothly in the presence of catalytic amounts of iron(II) ions.

$$H_2O_2 + Fe^{2+} \longrightarrow HO^- + HO\cdot + Fe^{3+} \;; \quad Fe^{3+} \xrightarrow{+e} Fe^{2+}$$

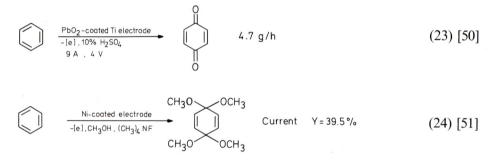

(21) [48]

Y = 64 %

The electrochemically generated silver ions oxidize benzene to quinone [49].

(22)

Y = 25~40 %

A titanium electrode coated with lead oxide or a nickel-coated electrode may be used for the electrochemical oxidation of benzene to quinone or quinone bisacetals, respectively.

4.7 g/h

(23) [50]

PbO$_2$-coated Ti electrode
-[e], 10% H$_2$SO$_4$
9 A , 4 V

Ni-coated electrode
-[e], CH$_3$OH, (CH$_3$)$_4$ NF

Current   Y = 39.5 %

(24) [51]

2.8.1.6 Acetamidation, Nitration, and Reactions with Pyridine

The anodic oxidation of aromatic compounds in the presence of acetonitrile leads to nuclear acetamidation.

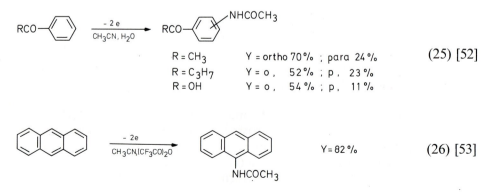

R = CH$_3$	Y = ortho 70% ; para 24%
R = C$_3$H$_7$	Y = o, 52% ; p, 23%
R = OH	Y = o, 54% ; p, 11%

(25) [52]

-2e
CH$_3$CN,(CF$_3$CO)$_2$O

Y = 82 %

(26) [53]

NHCOCH$_3$

In the absence of $(CF_3CO)_2O$ hydroxylation occurs. When the reaction site is blocked by a *t*-butyl group, dealkylative acetamidation takes place as shown below.

Y= 55% (27) [54]

The electrochemical oxidation of aromatic compounds in the presence of ammonium nitrate or N_2O_4 results in the nuclear nitration.

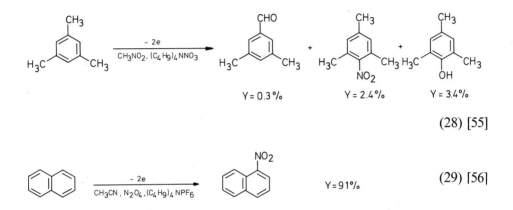

Y = 0.3%   Y = 2.4%   Y = 3.4%

(28) [55]

Y=91% (29) [56]

The cationic intermediates formed from the aromatic compounds react with pyridine at the nitrogen atom.

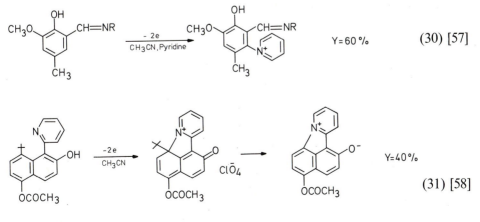

Y=60% (30) [57]

Y=40%

(31) [58]

97

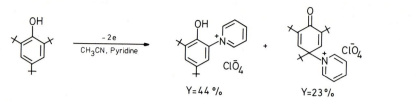

$$Y=44\% \qquad Y=23\%$$

(32) [59]

2.8.1.7 Halogenation

The mechanism of the nuclear halogenation depends on the relative values of the oxidation potentials of the halogen and aromatic substrate. When the oxidation potential of the halogen is lower than that of the aromatic compound, halogenation is initiated by the oxidation of halogen as it has already been shown in the previous section. Hence, this section is mainly concerned with the fluorination of aromatic compounds. One of the most important points in the anodic fluorination is the choice of the fluoride ion source. The use of anhydrous hydrofluoric acid usually leads to low yields [60–63]. Higher yields are obtained when a combination of tetraalkylammonium fluoride and hydrofluoric acid are employed, as shown in Table 6.

Table 6. Fluorination

Substrate	Solvent and supporting electrolyte	Product	Yield (%)	Ref.
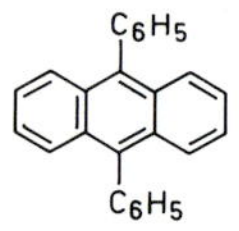 C_6H_5 / C_6H_5 (anthracene)	CH_3CN, $(C_2H_5)_4NF \cdot 3HF$	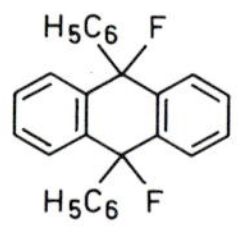 H_5C_6 F / H_5C_6 F	75	[64]
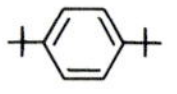 (di-t-butylbenzene)	CH_3CN, $(C_2H_5)_4NF \cdot 3HF$	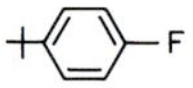 —F	63	[65]
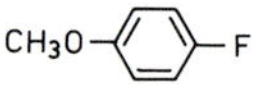 CH_3O——F	CH_3CN, $(C_2H_5)_4NF \cdot 3HF$	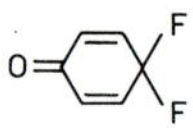 $O=$ ⟨ ⟩ $<^F_F$	50	[65]
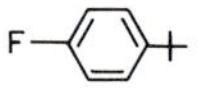 F—⟨ ⟩+	CH_3CN, $(C_2H_5)_4NF \cdot 3HF$	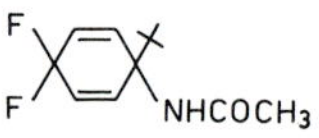 F, F $NHCOCH_3$	60	[65]
		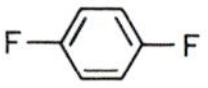 F—⟨ ⟩—F	5	
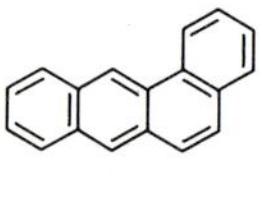 (benz[a]anthracene)	CH_3CN, $(CH_3)_4NF \cdot 2HF$	 (fluoro-benz[a]anthracene) F + two minor isomers	33	[66]

The chlorination of 9,10-diphenylanthracene is probably initiated by the oxidation of the aromatic system.

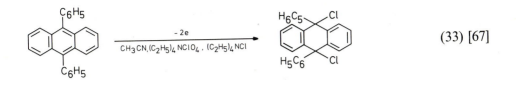

(33) [67]

Table 7. Cyanation

Substrate	Solvent and supporting electrolyte	Product	Yield (%)	Ref.
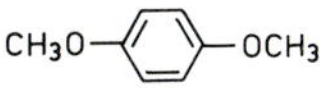 CH_3O—⟨⟩—OCH_3	CH_3CN, $(C_2H_5)_4NCN$	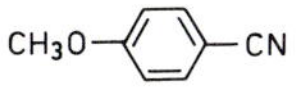 CH_3O—⟨⟩—CN	95^a	[69]
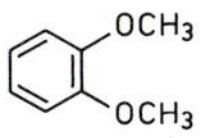 ⟨⟩ with OCH_3, OCH_3	CH_3CN, $(C_2H_5)_4NCN$	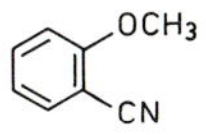 ⟨⟩ with OCH_3, CN	94^a	[69]
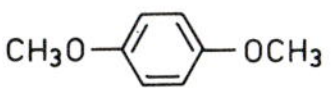 CH_3O—⟨⟩—OCH_3	CH_3OH, $NaCN$	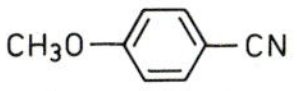 CH_3O—⟨⟩—CN + three products	33.5^a	[70]
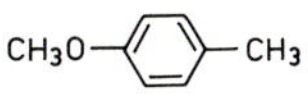 CH_3O—⟨⟩—CH_3	CH_3OH, $NaCN$	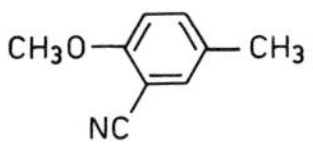 CH_3O—⟨⟩—CH_3 with NC + three products	17.4^a	[71]
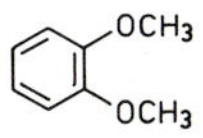 ⟨⟩ with OCH_3, OCH_3	H_2O, CH_2Cl_2, $NaCN$, $(C_4H_9)_4NSO_4H$	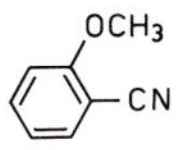 ⟨⟩ with OCH_3, CN	26	[72]
		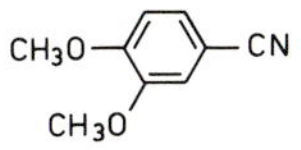 CH_3O—⟨⟩—CN with CH_3O	11	
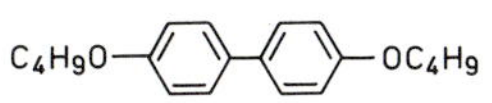 C_4H_9O—⟨⟩—⟨⟩—OC_4H_9	H_2O, CH_2Cl_2, $NaCN$, $(C_4H_9)_4NSO_4H$	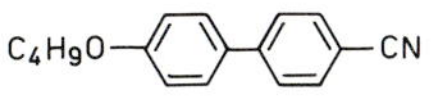 C_4H_9O—⟨⟩—⟨⟩—CN	66	[73]
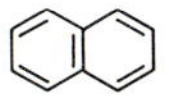 naphthalene	CH_3OH, $NaCN$	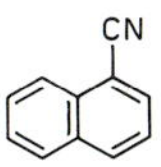 naphthalene with CN	74	[74]
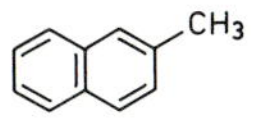 naphthalene with CH_3	CH_3OH, $NaCN$	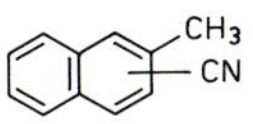 naphthalene with CH_3, CN	1—CN 60 4—CN 4 5—CN 6 8—CN 14	[74]

^{a} Current efficiency

2.8.1.8 Cyanation

The direct cyanation of the aromatic nucleus usually affords poor yields of cyanated products. However, considerably higher yields are obtained when one of the alkoxy groups of the starting alkoxyanisole or 4,4'-dialkoxy-biphenyl is substituted by a cyano group. Some examples of this type of substitution are given in Table 7.

2.8.2 Coupling

2.8.2.1 Intramolecular Coupling

When the structure of the substrate is suitable for coupling, the intramolecular coupling takes place rather easily.

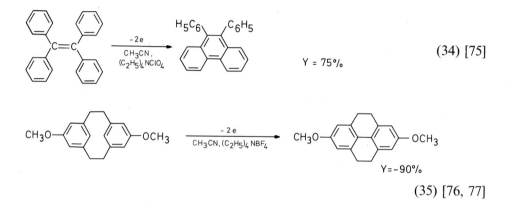

$$Y = 75\% \qquad (34) [75]$$

$$Y = \sim 90\% \qquad (35) [76, 77]$$

Coupling of the substrates $Ar(CH_2)_n Ar'$ is controlled by a variety of factors [78] including solvent and supporting electrolyte, length of n, molecular geometry, anode potential, and difference of the oxidation potentials of Ar and Ar'.

$$Ar(CH_2)_n Ar' \longrightarrow Ar \overset{(CH_2)_n}{\diagdown} Ar' \qquad (36)$$

The presence of trifluoroacetic acid or HBF_4 in the reaction system leads to satisfactory yields of coupled producrs [79–81].

Intramolecular coupling is readily achieved if n is 1, 2, 3, or 4 whereas with longer chain lengths intermolecular coupling predominates [82, 83]. A difference of 100 mV in the oxidation potentials of Ar and Ar' has been suggested to cause a strong effect on the reaction mechanism [84]. Since a variety of compounds have already been synthesized by using this intramolecular coupling, only some typical examples are given below.

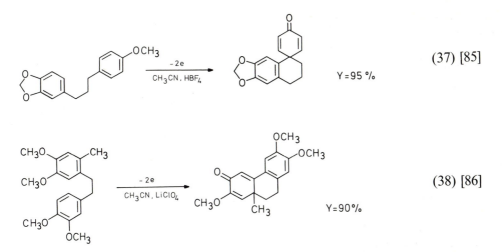

(37) [85]

Y = 95 %

(38) [86]

Y = 90%

The skeleton of morphine alkaroids has been synthesized by intramolecular coupling.

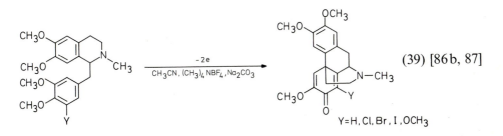

(39) [86 b, 87]

Y = H, Cl, Br, I, OCH₃

The product, (±)-O-methylflavinantine, is easily transformed to erybidine or glaucine [88].

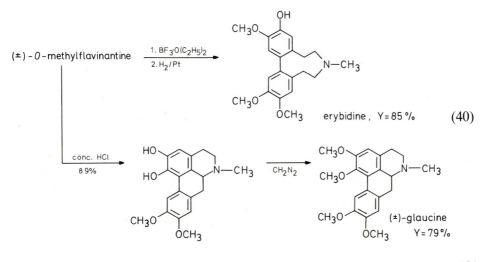

erybidine, Y = 85 %   (40)

(±)-glaucine Y = 79%

101

2. Anodic Oxidations

Substrates having the structure of $Ar-(CH_2)_n-X-(CH_2)_m-Ar'$ also undergo intramolecular coupling by anodic oxidation.

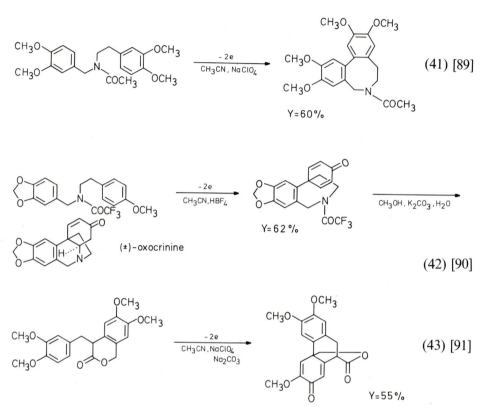

(41) [89]

Y=60%

(±)-oxocrinine

Y= 62 %

(42) [90]

(43) [91]

Y=55%

2.8.2.2 Intermolecular Coupling

When suitable aromatic compounds are oxidized in the absence of nucleophiles, the aromatic compounds themselves behave as nucleophiles to yield dimers. Table 8 shows some typical reactions.

Intermolecular cross-coupling has also been observed.

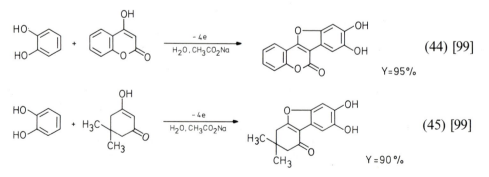

(44) [99]

Y=95%

(45) [99]

Y =90 %

Table 8. Intermolecular Coupling

Substrate	Solvent and supporting electrolyte	Product	Yield (%)	Ref.
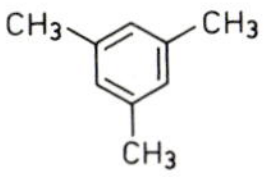 (1,3,5-trimethylbenzene)	CH_2Cl_2, $(C_4H_9)_4NBF_4$	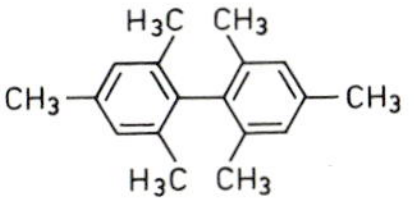 (coupled product)	53	[92]
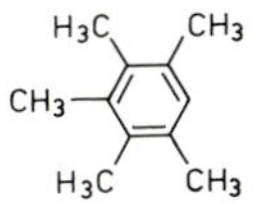 (pentamethylbenzene)	CH_2Cl_2, $(C_4H_9)_4NBF_4$	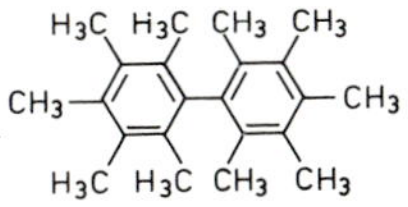 (coupled product)	30	[93]
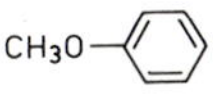 CH_3O–(anisole)	CH_2Cl_2, CF_3CO_2H, $(C_4H_9)_4NBF_4$	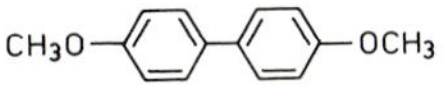 CH_3O––OCH_3	~100	[94] [95]
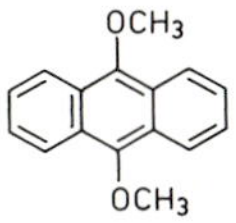 (9,10-dimethoxyanthracene)	CH_3CN, $LiClO_4$	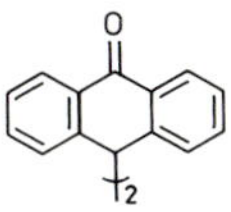 (anthraquinone deriv.)	~100	[96]
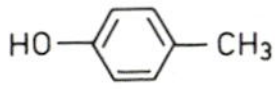 HO––CH_3	CH_3CN, H_2O, $NaClO_4$	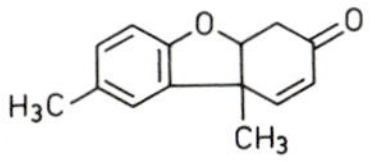 (spirocyclic product)	37	[97]
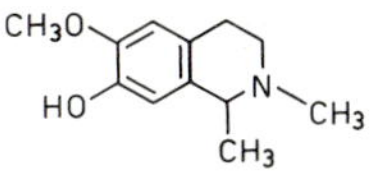 (methoxy-hydroxy tetrahydroisoquinoline)	CH_3CN, CH_3ONa	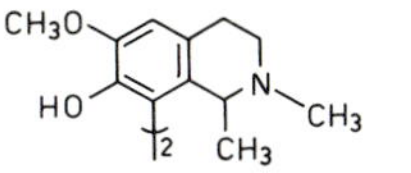 (dimeric product)	69	[98]

2.8.3 Oxidation at the Benzylic Position

As described in the previous chapter, substitution at the benzylic position always takes place together with nuclear substitution if the aromatic substrates possess benzylic hydrogens [100].

$$ArCH_2R \xrightarrow{-e} \overset{+\cdot}{Ar}CH_2R \xrightarrow{-H^+} Ar\dot{C}HR \xrightarrow{-e} Ar\overset{+}{C}HR \xrightarrow{Nu^-} Ar\underset{Nu}{C}HR \qquad (46)$$

Since the mechanism of the benzylic substitution has already been reviewed in detail [100–102], only some reactions which are interesting from the standpoint of organic synthesis are described below.

103

2. Anodic Oxidations

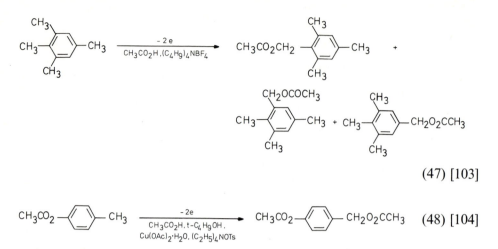

(47) [103]

$$CH_3CO_2 - \underset{}{\bigcirc} - CH_3 \xrightarrow[\substack{CH_3CO_2H.\,t-C_4H_9OH. \\ Cu(OAc)_2\cdot H_2O,\ (C_2H_5)_4NOTs}]{-2e} CH_3CO_2 - \underset{}{\bigcirc} - CH_2O_2CCH_3 \quad (48)\ [104]$$

Modification of the steroid skeleton seems to be particularly interesting.

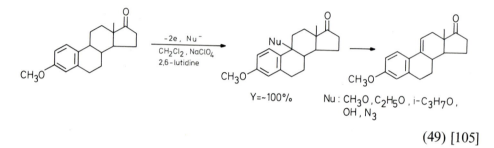

(49) [105]

Benzylic alcohols, esters, and ethers can also be oxidized at the benzylic position to yield the corresponding carbonyl compounds [106–111]. The proposed mechanism for the cleavage of benzylic ethers is similar to benzylic substitution [108, 112].

$$ArCH_2OR \xrightarrow{-e} \overset{+\cdot}{Ar}CH_2OR \xrightarrow{-H^+} Ar\overset{\cdot}{C}HOR \xrightarrow{-e} Ar\overset{+}{C}HOR \xrightarrow{H_2O} Ar\overset{\overset{\displaystyle +}{O}H_2}{\underset{|}{C}HOR} \quad (50)$$

$$\xrightarrow{-H^+} ArCHO \ + \ ROH$$

Carbon—carbon bond cleavage can also take place at the benzylic position, if the intermediate cationic species is sufficiently stabilized by suitable substituents.

$$ArCH_2CH_2Y \xrightarrow{-2e} ArCH_2^+ + {}^+CH_2Y \xrightarrow{Nu^-} ArCH_2Nu + YCH_2Nu \quad (51)$$

Y : cation stabilizing groups

104

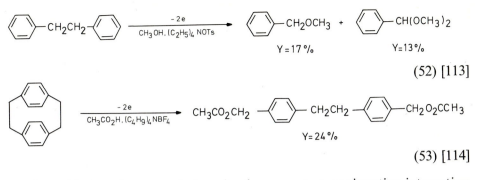

$$Y = 17\%$$

$$Y = 13\%$$

(52) [113]

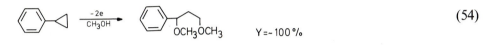

$$Y = 24\%$$

(53) [114]

In the oxidative cleavage of phenylcyclopropane, a conjugative interaction between the aromatic π-electron system and the cyclopropane ring has been suggested [115].

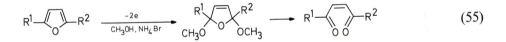

$$Y = \sim 100\%$$

(54)

2.8.4 Oxidation of Furan, Pyrrole, and Thiophene

2.8.4.1 Furans

The anodic oxidation of furans is one of the most extensively studied reactions because electrooxidation of furans in methanol yields 2,5-dimethoxy-2,5-dihydrofurans which are useful starting materials in organic synthesis [116–118].

(55)

2,5-Dimethoxy-2,5-dihydrofuran derivatives have been extensively utilized for the synthesis of aromatic and aliphatic ring systems.

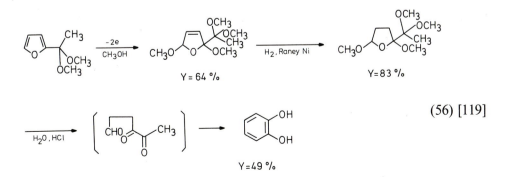

$$Y = 64\%$$

$$Y = 83\%$$

(56) [119]

$$Y = 49\%$$

105

2. Anodic Oxidations

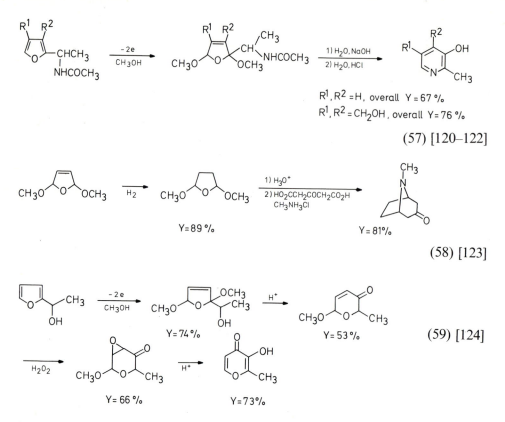

$R^1, R^2 =$ H, overall Y = 67 %

$R^1, R^2 =$ CH$_2$OH, overall Y = 76 %

(57) [120–122]

Y = 89 %   Y = 81%

(58) [123]

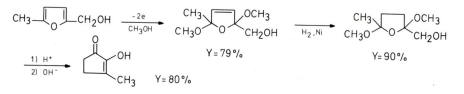

Y = 74 %   Y = 53 %   (59) [124]

Y = 66 %   Y = 73%

A variety of cyclopentenone derivatives have been synthesized using the anodic oxidation of furans as a key step.

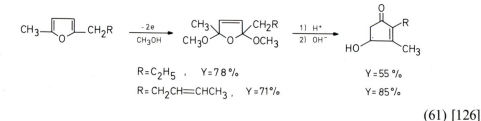

Y = 79 %   Y = 90%

Y = 80%

(60) [125]

R = C$_2$H$_5$,   Y = 78 %   Y = 55 %

R = CH$_2$CH=CHCH$_3$,   Y = 71%   Y = 85 %

(61) [126]

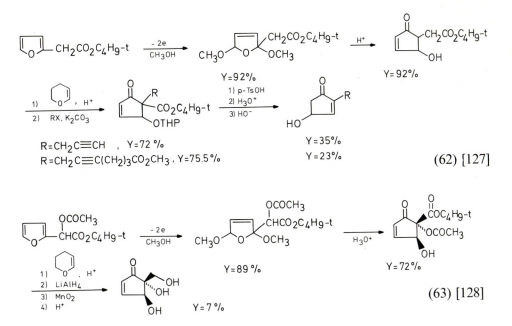

$R=CH_2C\equiv CH$, $Y=72\%$
$R=CH_2C\equiv C(CH_2)_3CO_2CH_3$, $Y=75.5\%$

(62) [127]

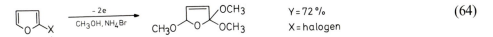

(63) [128]

The formation of 2,2,5-trimethoxy-2,5-dihydrofuran has been reported [129].

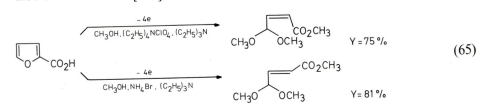

(64)

Ring opening involving formation of the trimethoxylated compound has also been observed [130].

(65)

Besides methoxylation, acetoxylation and hydroxylation of furan may also be achieved.

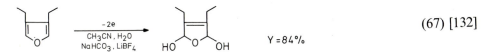

(66) [131]

(67) [132]

107

2. Anodic Oxidations

The elimination of one molecule of acetic acid from 2,5-diacetoxy-2,5-dihydrofuran yields 2-acetoxyfuran which can react with electrophiles to yield butenolide derivatives.

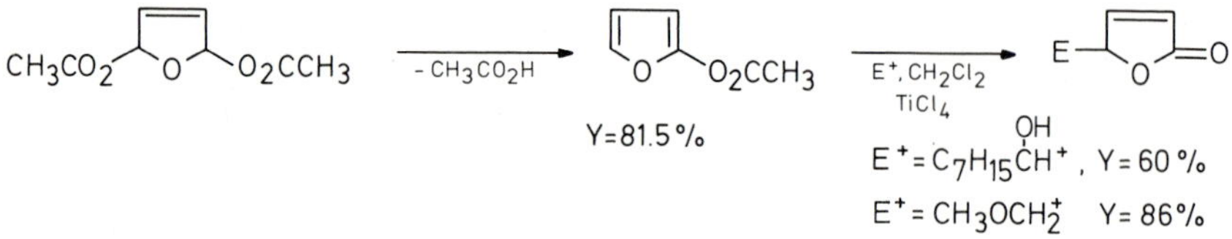

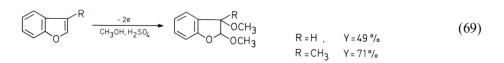

$$Y=81.5\%$$

$$E^+ = C_7H_{15}\overset{OH}{CH^+}, \quad Y=60\%$$

$$E^+ = CH_3OCH_2^+ \quad Y=86\%$$

(68) [131, 133]

The oxidation of benzofuran has also been reported [134].

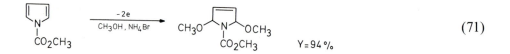

R = H ,   Y = 49 %
R = CH₃   Y = 71%

(69)

2.8.4.2 Pyrrole and Thiophene

The anodic oxidation of 1-methylpyrrole proceeding via the intermediate 2,5-dimethoxylated product yields 1-methyl-2,2,5,5-tetramethoxy-2,5-dihydropyrrole [135, 136].

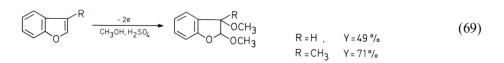

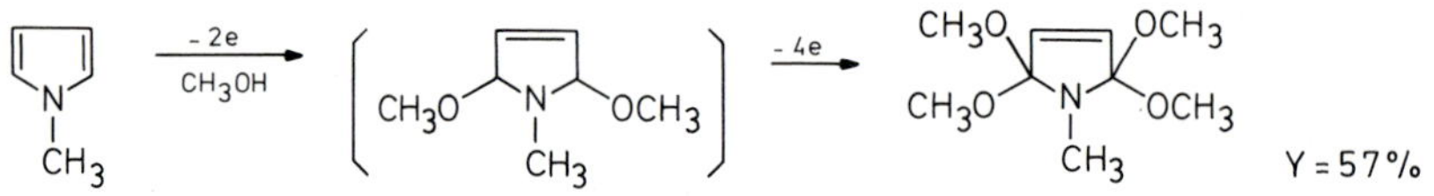

(70)

On the other hand, the electrochemical oxidation of pyrrole itself yields polymeric products [133, 137]. However, 1-methoxycarbonyl pyrrole is oxidized to 2,5-dimethoxy-1-methoxycarbonyl-2,5-dihydropyrrole in high yield [138].

(71)

When the reaction is carried out in acetic acid/sodium acetate, 2,5-diacetoxy-1-methoxycarbonyl-2,5-dihydropyrrole is obtained:

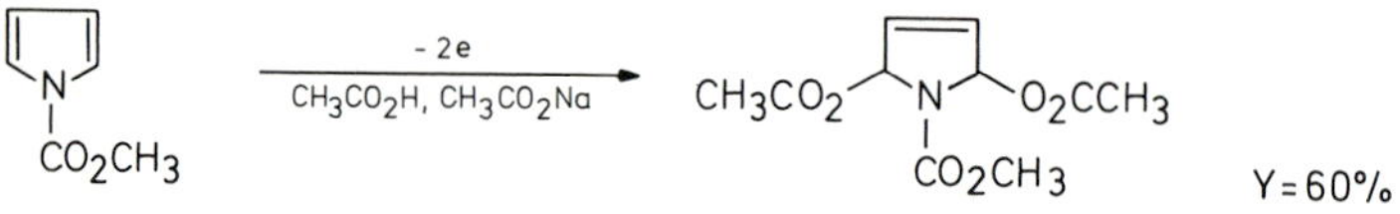

(72)

Cyanation of pyrrole derivatives yields the expected products:

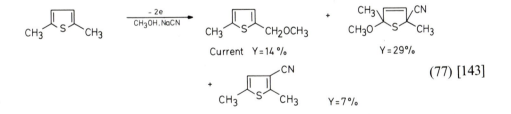

Thiophene does not show satisfactory selectivity in cyanation:

References

1. Linstead, R. P., Bunt, J. C., Weedon, B. C. L., Shephand, B. R.: J. Chem. Soc. *1952*, 3625
2. Eberson, L., Nyberg, K.: Acc. Chem. Res. *6*, 106 (1973)
3. Leung, M., Herz, J., Salzberg, H. W.: J. Org. Chem. *30*, 310 (1965)
4. Eberson, L., Nyberg, K.: J. Am. Chem. Soc. *88*, 1686 (1966)
5. Ross, S. D., Finkelstein, M., Peterson, R. C.: J. Am. Chem. Soc. *89*, 4088 (1967)
6. Yoshida, K., Shigu, M., Kanbe, T., Fueno, T.: J. Org. Chem. *40*, 3805 (1975)
7. Nyberg, K., Wistrand, L. G.: J. Chem. Soc., Chem. Commun. *1976*, 898

8. So, Y. H., Miller, L. L.: Synthesis *1976*, 468
9. So, Y. H., Becker, J. Y., Miller, L. L.: J. Chem. Soc. *1975*, 262
10. Weinberg, N. L., Wu, C. N.: Tetrahedron Lett. *1975*, 3367
11. Bockmair, G., Fritz, H. P., Gebauer, H.: Electrochim. Acta *23*, 21 (1978)
12. Blum, Z., Cedheim, L., Nyberg, K.: Acta Chem. Scand. *B29*, 715 (1975)
13. Eberson, L., Oberrauch, E.: Acta Chem. Scand. *B35*, 193 (1981)
14. Bockmain, G., Fritz, H. P.: Electrochim. Acta *21*, 1099 (1976)
15. Parker, V. D., Dirlan, J. P., Eberson, L.: Acta Chem. Scand. *B25*, 341 (1971)
16. Belleau, B., Weinberg, N. L.: J. Am. Chem. Soc. *85*, 2525 (1963)
17. Weinberg, N. L., Belleau, B.: Tetrahedron *29*, 279 (1973)
18. Margaretha, P., Tissot, P.: Helv. Chim. Acta *58*, 103 (1975)
19. Manning, M. J., Raynolds, P. W., Swenton, J. S.: J. Am. Chem. Soc. *98*, 5008 (1976)
20. a) Chenard, B. L., Swenton, J. S.: J. Chem. Soc., Chem. Commun. *1979*, 1172
    b) See also Margaretha, P.: Helv. Chim. Acta *62*, 1978 (1979)
21. Manning, M. J., Henton, D. R., Swenton, J. S.: Tetrahedron Lett. *1977*, 1679
22. Dolson, M. G., Swenton, J. S.: J. Am. Chem. Soc. *103*, 2361 (1981)
23. Swenton, J. S., Raynolds, P. W.: J. Am. Chem. Soc. *100*, 6188 (1978)
24. Buchanan, G. L., Raphael, R. A., Taylor, R.: J. Chem. Soc., Perkin Trans. (1) *1973*, 373
25. Evans, D. A., Tanis, S. P., Hart, D. J.: J. Am. Chem. Soc. *103*, 5813 (1981)
26. Bornatsch, W., Vogel, E.: Angew. Chem. *87*, 412 (1975)
27. Nilsson, A., Ronlán, A.: Tetrahedron Lett. *1975*, 1107
28. Nilsson, A., Palmquist, U., Petterson, T., Ronlán, A.: J. Chem. Soc. Perkin Trans. (1) *1978*, 696
29. Stewart, R. F., Miller, L. L.: J. Am. Chem. Soc. *102*, 4999 (1980)
30. Henton, D. R., Anderson, K., Manning, M. J., Swenton, J. S.: J. Org. Chem. *45*, 3422 (1980)
31. Groebel, B. T., Konz, E., Millauer, H., Pistorius, R.: Synthesis *1979*, 605
32. Dolson, M. G., Swenton, J. S.: J. Org. Chem. *46*, 177 (1981)
33. Cowitz, F. H.: US Patent 3509039 (1970); C.A. *73*, 115824 (1970)
34. Jones, G. C., Payne, O. A.: US Patent 3994788 (1976); C.A. *86*, 35759 (1977)
35. Bhat, G. A., Periasamy, M., Bhatt, M. V.: Tetrahedron Lett. *1979*, 3097
36. Brinker, U. H., Tyner, M., III, Jones, W. M.: Synthesis *1975*, 671
37. Blumn, Z., Nyberg, K.: Acta Chem. Scand. *B33*, 73 (1979)
38. Parker, V. D.: Acta Chem. Scand. *B24*, 2775 (1970)
39. Brachi, J., Rieker, A.: Synthesis *1977*, 708
40. Johnson, R. W., Grover, E. R., MacPherson, L..J.: Tetrahedron Lett. *22*, 3719 (1981)
41. Johnson, R. W., Bendnarski, M. D., O'Leary, B. F., Grover, E. R.: Tetrahedron Lett. *22*, 3715 (1981)
42. Utley, J. H. P., Yates, J. B.: J. Chem. Soc., Chem. Commun. *1973*, 473
43. Nillson, A., Ronlan, A., Parker, V. D.: J. Chem. Soc., Perkin Trans. (1), *1973*, 2337
44. Vermillion, F. J., Jr., Pearl, I. A.: J. Electrochem. Soc. *111*, 1392 (1964)
45. Rieker, A., Treher, E. L., Giesel, H., Khalifa, M. H.: Synthesis *1978*, 851
46. Masui, M., Ueda, C., Ohmori, H.: Chem. Pharm. Bull. *26*, 1953 (1978)
47. Blumn, Z., Cedheim, L., Nyberg, K.: Acta Chem. Scand. *B29*, 715 (1975)
48. a) Steckhan, E., Wellmann, J.: Angew. Chem. *88*, 306 (1976)
    b) Wellmann, J., Steckhan, E.: Chem. Ber. *110*, 3561 (1977)

49. a) Fleischmann, M., Pletcher, D., Rafinski, A.: J. Appl. Electrochem. *1*, 1 (1971)
    b) Goodridge, F., Umeh, E. O.: Electrochim. Acta, *20*, 991 (1975)
50. Millington, J. P.: Brit. Patent 1377681 (1974); C.A. *82*, 125099 (1975)
51. a) Cramer, J.: Ger. Offen. 2739315 (1979); C.A. *90*, 186593 (1979)
    b) Cramer, J.: Ger. Offen. 2739316 (1979); C.A. *90*, 186593 (1979)
52. So, Y. H., Becker, J. Y., Miller, L. L.: J. Chem. Soc., Chem. Commun. *1975*, 262
53. Hammerich, O., Parker, V. D.: J. Chem. Soc., Chem. Commun. *1974*, 245
54. Dreher, E. L., Bracht, J., El-Mobayed, M., Hütter, P., Winter, W., Rieker, A.: Chem. Ber. *115*, 288 (1982)
55. Nyberg, K.: Acta Chem. Scand. *25*, 3246 (1971)
56. a) Achord, J. M., Hussey, C. L.: J. Electrochem. Soc. *128*, 2556 (1981)
    See also b) Eberson, L., Jönsson, L., Radner, F.: Acra Chem. Scand. *B32*, 749 (1978)
56. c) Eberson, L., Radner, F.: Acta Chem. Scand. *B34*, 739 (1980)
    d) Perrin, C. L.: J. Am. Chem. Soc. *99*, 5516 (1977)
57. Ohmori, H., Matsumoto, A., Masui, M.: J. Chem. Soc., Chem. Commun. *1978*, 407
58. Popp, G.: J. Org. Chem. *37*, 3058 (1972)
59. Popp, G., Reitz, N. C.: J. Org. Chem. *37*, 3646 (1972)
60. Huba, F., Yaeger, E. B., Olah, G. A.: Electrochim. Acta *24*, 489 (1979)
61. Davis, V. J., Haszeldine, R. N., Tipping, A. E.: J. Chem. Soc., Perkin Trans. (1), *1975*, 1263
62. Inoue, Y., Nagase, S., Kodaira, K., Baba, H., Abe, T.: Bull. Chem. Soc. Jpn. *46*, 2204 (1973)
63. Yonekura, M., Nagase, S., Baba, H., Kodaira, K., Abe, T.: Bull. Chem. Soc. Jpn. *49*, 1113 (1976)
64. a) Rozhkov, I. N., Gambaryan, N. P., Galpern, E. G.: Tetrahedron Lett. *1976*, 4819
    b) Ludman, C. J., McCarron, E. M., O'Malley, R. F.: J. Electrochem. Soc. *119*, 874 (1972)
65. Rozhkov, I. N., Alyev, I. Y.: Tetrahedron *31*, 977 (1975)
66. O'Malley, R. F., Mariani, H. A.: J. Org. Chem. *46*, 2816 (1981)
67. Evans, J. F., Blount, H. N.: J. Org. Chem. *41*, 516 (1976)
68. a) Koyama, K., Susuki, T., Tsutsumi, S.: Tetrahedron Lett. *1965*, 627
    b) Koyama, K., Susuki, T., Tsutsumi, S.: Tetrahedron *23*, 2675 (1966)
69. Andreades, S., Zahnow, E. W.: J. Am. Chem. Soc. *91*, 4181 (1969)
70. Weinberg, N. L., Marr, D. H., Wu, C. N.: J. Am. Chem. Soc. *97*, 1499 (1975)
71. Yoshida, K., Shigi, M., Fueno, T.: J. Org. Chem. *40*, 63 (1975)
72. Eberson, L., Helgée, B.: Acta Chem. Scand. *B29*, 451 (1975)
73. Eberson, L., Helgée, B.: Acta Chem. Scand. *B31*, 813 (1977)
74. Yoshida, K., Nagase, S.: J. Am. Chem. Soc. *101*, 4268 (1979)
75. Stuart, J. D., Ohnesorge, W. E.: J. Am. Chem. Soc. *93*, 4531 (1971)
76. Becker, J. L., Miller, L. L., Boekelheide, V., Morgan, T.: Tetrahedron Lett. *1976*, 2939
77. Kerr, J. B., Jempty, T. C., Miller, L. L.: J. Am. Chem. Soc. *101*, 7338 (1979)
78. Parker, V. D., Ronlán, A.: J. Am. Chem. Soc. *97*, 4717 (1975)
79. Bechgaard, K., Hammerich, O., Moe, N. S., Ronlán, A., Svanholm, U., Parker, V. D.: Tetrahedron Lett. *1972*, 2271
80. Kotani, E., Tobinaga, S.: Tetrahedron Lett. *1973*, 4759 (1973)
81. Ronlán, A., Hammerich, O., Parker, V. D.: J. Am. Chem. Soc. *95*, 7132 (1973)

82. Ronlán, A., Parker, V. D.: J. Org. Chem. *39*, 1014 (1974)
83. Nilsson, A., Palmquist, U., Ronlán, A., Parker, V. D.: J. Am. Chem. Soc. *97*, 3540 (1975)
84. Palmquist, U., Nilsson, A., Parker, V. D., Ronlán, A.: J. Am. Chem. Soc. *98*, 2571 (1976)
85. a) Kotani, E., Takeuchi, N., Tobinaga, S.: J. Chem. Soc., Chem. Commun. *1973*, 550
    See also b) Schwartz, M. A., Rose, B. F., Holton, R. A., Scott, S. W., Vishmuvajjale, B.: J. Am. Chem. Soc. *99*, 2571 (1977)
    c) Palquist, U., Nilsson, A., Pettersson, T., Ronlán, A.: J. Org. Chem. *44*, 196 (1979)
    d) Palmquist, U., Ronlán, A., Parker, V. D.: Acta Chem. Scand. *B28*, 264 (1974)
    e) Parker, V. D., Palmquist, U., Ronlán, A.: Acta Chem. Scand. *B28*, 1241 (1974)
86. a) Falck, J. R., Miller, L. L., Stermitz, F. R.: J. Am. Chem. Soc. *96*, 2981 (1974)
    b) Miller, L. L., Stewart, R. F.: J. Org. Chem. *43*, 1580 (1978)
87. a) Miller, L. L., Stermitz, F. R., Falck, J. R.: J. Am. Chem. Soc. *93*, 5941 (1971)
    b) Miller, L. L., Stermitz, F. R., Falck, J. R.: J. Am. Chem. Soc. *95*, 2651 (1973)
    c) Falck, J. R., Miller, L. L., Stermitz, F. R.: Tetrahedron *30*, 931 (1974)
    d) Miller, L. L., Stermitz, F. R., Becker, J. Y., Ramachandran, V.: J. Am. Chem. Soc. *97*, 2922 (1975)
    See also e) ref. [6]
    f) Bobbitt, J. M., Noguchi, J., Ware, R. S., Ching, K. N., Huang, S. J.: J. Org. Chem. *40*, 2924 (1975)
88. Kupchan, S. M., Kim, C.: J. Am. Chem. Soc. *97*, 5623 (1975)
89. a) Sainsbury, M., Wyatt, J.: J. Chem. Soc., Perkin Trans. (1), *1976*, 661
    b) Sainsbury, M., Wyatt, J.: J. Chem. Soc., Perkin Trans. (1), *1977*, 1750
90. a) See ref. [11.a)]
    b) See also Elliott, I. W., Jr.: J. Org. Chem. *42*, 1090 (1977)
91. a) Sainsbury, M., Schinazi, R. F.: J. Chem. Soc., Chem. Commun. *1972*, 718
    b) See also Palmquist, U., Nilsson, A., Pettersson, T., Ronlán, A.: J. Org. Chem. *44*, 196 (1979)
92. a) Nyberg, K.: Acta Chem. Scand. *B24*, 1609 (1970)
    b) Nyberg, K.: Acta Chem. Scand. *B25*, 534 (1971)
    See also c) Majeski, E. J., Stuart, J. D., Ohnesorge, W. E.: J. Am. Chem. Soc., *90*, 633 (1968)
    d) Nyberg, K.: Acta Chem. Scand. *B24*, 2757, 3151, 3162, 3171 (1970)
93. Nyberg, K.: Acta Chem. Scand. *B25*, 2499 (1971)
94. Ronlán, A., Beckgaard, K., Parker, V. D.: Acta Chem. Scand. *B27*, 2375 (1973)
95. a) Beckgaard, K., Hammerich, D., Moe, M. S., Ronlán, A., Svanholm, U., Parker, V. D.: Tetrahedron Lett. *1972*, 2271
    b) See also Nilsson, A., Palmquist, U., Petterson, T., Ronlán, A.: J. Chem. Soc., Perkin Trans. (1), *1978*, 696
96. Parker, V. D.: Tetrahedron Lett. *1972*, 1449
97. Miller, L. L., Stewart, R. F.: J. Org. Chem. *43*, 1580 (1978)
98. a) Bobbitt, J. M., Noguchi, I., Yagi, H., Weisgraber, K. H.: J. Am. Chem. Soc., *93*, 3552 (1971)

   b) Bobbitt, J. M., Noguchi, I., Yagi, H., Weisgraber, K. H.: J. Org. Chem. *41*, 845 (1976)

   c) Bobbitt, J. M., Yagi, H., Shibuya, S., Stock, J. T.: J. Org. Chem. *36*, 3006 (1971)

99. Grujić, Z., Tabaković, I., Trkovik, M.: Tetrahedron Lett. *1976*, 4823
100. Eberson, L., Nyberg, K.: Acc. Chem. Res. *6*, 106 (1973)
101. Eberson, L., Sternerup, H.: Acta Chem. Scand. *B26*, 1431, 1454 (1972)
102. Sternerup, H.: Acta Chem. Scand. *B28*, 969 (1974)
103. Eberson, L., Oberrauch, E.: Acta Chem. Scand. *B33*, 343 (1979)
104. Torii, S., Tanaka, H., Siroi, T., Akada, M.: J. Org. Chem. *44*, 3305 (1979)
105. Ronsold, K., Kash, H.: Tetrahedron Lett. *1979*, 4463
106. Miller, L. L., Koch, V. R., Larscheid, M. E., Wolf, J. F.: Tetrahedron Lett. *1971*, 1389
107. Miller, L. L., Wolf, J. F., Mayeda, E. A.: J. Am. Chem. Soc. *93*, 3306 (1971)
108. Mayeda, E. A., Miller, L. L., Wolf, J. F.: J. Am. Chem. Soc. *94*, 6812 (1972)
109. Garwood, R. F., Dim, N. ud., Weedon, B. C. L.: J. Chem. Soc., Perkin Trans. (1), *1975*, 2471
110. Weinreb, S. M., Epling, G. A., Comi, R., Reitano, M.: J. Org. Chem. *40*, 1356 (1975)
111. Lines, R., Utley, J. H. P.: J. Chem. Soc., Perkin Trans. (2), *1977*, 803
112. Boyd, J. W., Schmalzl, P. W., Miller, L. L.: J. Am. Chem. Soc. *102*, 3856 (1980)
113. Shono, T., Matsumura, Y.: Japan Chemical Society 3rd Symposium on Oxidation, Abstract, p. 101, 1969
114. Sato, T., Torizuka, K., Shimizu, M., Kurihara, Y., Yoda, N.: Bull. Chem. Soc. Jpn. *52*, 2420 (1979)
115. Shono, T., Matsumura, Y.: J. Org. Chem. *35*, 4157 (1970)
116. Kaas, N. C., Limborg, F., Glens, K.: Acta Chem. Scand. *6*, 531 (1952)
117. Elming, N.: Adv. Org. Chem. *2*, 67 (1960), and references cited therein
118. Weinberg, N. L., Weinberg, H. R.: Chem. Rev. *1968*, 449
119. Nielsen, J. T., Elming, N., Clauson-Kaas, N.: Acta Chem. Scand. *9*, 9 (1955)
120. Kaas, N. C., Elming, N., Tyle, Z.: Acta Chem. Scand. *9*, 1 (1955)
121. Elming, N., Kaas, N. C.: Acta Chem. Scand. *9*, 23 (1955)
122. Shono, T., Matsumura, Y., Tsubata, K., Takata, J.: Chem. Lett. *1981*, 1121
123. Elming, N., Nedenskov, P.: Brit. Patent 791770 (1958)
124. Shono, T., Matsumura, Y.: Tetrahedron Lett. *1976*, 1363
125. Shono, T., Matsumura, Y., Hamaguchi, H.: J. Chem. Soc., Chem. Commun. *1977*, 712
126. Shono, T., Matsumura, Y., Hamaguchi, H., Nakamura, K.: Chem. Lett. *1976*, 1249
127. Shono, T., Hamaguchi, H., Aoki, K.: Chem. Lett. *1977*, 1053
128. Shono, T., Matsumura, Y., Yamane, S., Suzuki, M.: Chem. Lett. *1980*, 1619
129. Hillers, S., Sokolov, G. P., Karmilćhik, A. Ya.: C.A. *62*, 6449 (1965)
130. Tanaka, H., Kobayashi, Y., Torii, S.: J. Org. Chem. *41*, 3482 (1976)
131. a) Shono, T., Matsumura, Y., Yamane, S.: Tetrahedron Lett. *22*, 3269 (1981)
    See also b) Baggaley, A. J., Brettle, R.: J. Chem. Soc. (C) *1968*, 969
   c) Kolb, K. E.: J. Chem. Soc., Chem. Commun. *1966*, 271
132. Froborg, J., Magnusson, G., Thórén, S.: J. Org. Chem. *40*, 122 (1975)
133. Cava, M. P., Wilson, C. L., Williams, C. J., Jr.: J. Am. Chem. Soc. *78*, 2303 (1956)

134. Janda, L. M., Stibor, I., Rozinek, R.: Synthesis *1975*, 717
135. Gabel, N. W.: J. Org. Chem. *27*, 301 (1962)
136. Weinberg, N. L., Brown, F. A.: J. Org. Chem. *31*, 4054 (1966)
137. Diaz, A., Kanazawa, K., Gardini, G. P.: J. Chem. Soc., Chem. Commun. *1970*, 635
138. Shono, T., Matsumura, Y., Yamane, S., Takata, T.: The 45th Annual Meeting of the Chemical Society of Japan (1982), Abstract, p. 1029
139. Eberson, L.: Acta Chem. Scand. *B34*, 747 (1980)
140. Yoshida, K.: J. Am. Chem. Soc. *99*, 6111 (1977)
141. Yoshida, K.: J. Am. Chem. Soc. *101*, 2116 (1979)
142. Torii, S., Yamanaka, T., Tanaka, H.: J. Org. Chem. *43*, 2882 (1978)
143. Yoshida, K., Saeki, T., Fueno, T.: J. Org. Chem. *36*, 3673 (1971)

2.9 Oxidations Using Mediators

2.9.1 Principles

As described in the previous parts of this book, the active species are generally generated by direct electron transfer between substrate and electrode in the electroorganic reactions. Hence, the formation of active species is highly controlled by the oxidation and reduction potentials of the substrates. When these potentials are beyond the range accessible by the usual electrochemical technique, the direct electron transfer between the substrate and electrode hardly takes place as described in the direct oxidation of aliphatic saturated alcohols. Therefore, it is necessary to devise some other methods to oxidize or reduce the substrates. Also, even if the oxidation and reduction potentials of the substrates are in the accessible range of the electrochemical method, it is more desirable to oxidize or reduce them at much lower potentials than those applied in the direct method. This is achieved by the electroorganic synthesis using mediators. The oxidative reaction system using a mediator, which is called a mediatory system, is schematically represented in Fig. 1.

The oxidation potential of the substrate S in Fig. 1 is beyond the range accessible by the electrochemical method so that direct electron transfer

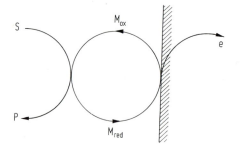

Fig. 1. Mediatory System

from S to the anode hardly occurs, and also the high oxidation potential necessary for the direct oxidation of S causes unexpected side reactions involving oxidation of the solvent or supporting electrolyte. However, when a compound M_{red} (a reduced form of M) which may be oxidized at sufficiently lower potential than S is added to the reaction system, the oxidation of M_{red} to M_{ox} (an oxidized form of M) will take place prior to the oxidation of S. Provided that M_{ox} is able to oxidize S to product P, the oxidation of S will be achieved at a potential lower than that necessary for its direct oxidation. Oxidation of S with M_{ox} may be effected in two ways, namely by direct electron transfer (homogeneous electron transfer) from S to M_{ox} in solution or by chemical oxidation of S with M_{ox}. In this book the former system is called homomediatory system and the latter heteromediatory (or chemomediatory) system. The compound M is called a mediator or an electron carrier since M mediates electron transfer between S and anode. When M_{ox} oxidizes S in solution, M_{ox} is reduced to M_{red} which is again oxidized at the anode to regenerate M_{ox}. Thus, if the life time of the redox system $M_{ox} \rightleftarrows M_{red}$ is sufficiently long, only a catalytic amount of the mediator is required to initiate the entire reaction. As a matter of course, the concept of the mediatory system is not only applicable to oxidations as illustrated by Fig. 1 but also to reductions. Although the term mediator or electron carrier has been introduced rather recently, many types of reaction systems involving a compound which behaves as a mediator had already been known.

2.9.2 Homomediatory Systems

The homomediatory system is represented by Eqs. (1a) to (1c), in which the mediator M is first oxidized to the cation radical $M^{+\cdot}$ at a relatively low oxidation potential. The next step involves a homogeneous electron transfer from S to $M^{+\cdot}$ to form $S^{+\cdot}$; this step is a reversible reaction. In the final step, $S^{+\cdot}$ is transformed to the products P_1^+ and $P_2^{\cdot}$ by an irreversible reaction.

$$M \xrightarrow{-e} M^{+\cdot} \tag{1a}$$

$$M^{+\cdot} + S \rightleftarrows M + S^{+\cdot} \tag{1b}$$

$$S^{+\cdot} \longrightarrow P_1^+ + P_2^{\cdot} \tag{1c}$$

Since the oxidation potential of S is more positive than that of M, the equilibrium (1b) is largely shifted to the left side. Hence, the rate of the whole reaction greatly depends on the rate of the irreversible reaction (1c). In fact, the oxidation described by Eq. (1) proceeds effectively only when $S^{+\cdot}$ is transformed sufficiently fast to products P_1^+ and $P_2^{\cdot}$.

When the oxidation potential of S is extremely more positive than M, the oxidation illustrated by Eq. (1) is almost imposible, even though the irre-

2. Anodic Oxidations

versible reaction of $S^{+\cdot}$ is fast. In such a case, some further activation of $M^{+\cdot}$ is necessary to make the oxidation possible.

The anodic oxidation of carboxylate anions in the presence of *tris*-(*p*-bromophenyl)amine (M) is one of the earliest typical reactions using the homomediatory system (2) [1].

$$(Br-\!\!\!\bigcirc\!\!\!-)_3N \xrightarrow{-e} (Br-\!\!\!\bigcirc\!\!\!-)_3\overset{+\cdot}{N}$$

$$(Br-\!\!\!\bigcirc\!\!\!-)_3\overset{+\cdot}{N} + RCO_2^- \rightleftharpoons (Br-\!\!\!\bigcirc\!\!\!-)_3N + RCO_2{}^\cdot \tag{2}$$

$$RCO_2{}^\cdot \longrightarrow R\cdot + CO_2$$

Since the oxidation half-wave potential of M (1.30 V *vs.* NHE) is sufficiently low so that M may be oxidized prior to oxidation of the carboxylate anion and the collapse of the acyloxyl radical is very fast (half-life about 10^{-10}/s), the oxidation of carboxylate anions may be promoted by the mediator. Some other studies are summarized in Table 1.

Table 1. Homomediatory Oxidation

Substrate	Product	Yield (%)	Mediator	Ref.
$C_8H_{17}OCH_2-\!\!\bigcirc\!\!-OCH_3$	$C_8H_{17}OH$	95	$(Br-\!\!\bigcirc\!\!-)_3N$	[2]
$CH_2=CH-CH=CH_2$	$BrCH_2CHCH=CH_2$ $\;\;\;\;\;\;\overset{|}{O}CH_3$	97	$(Br-\!\!\bigcirc\!\!-)_3N$	[3]
(dithiane structure)	$\bigcirc\!\!=\!\!O$	95	$(CH_3-\!\!\bigcirc\!\!-)_3N$	[4]
$\underset{C_2H_5O_2C}{\overset{C_8H_{17}}{>}}\!\!<\!\!\overset{S}{\underset{S}{\rangle}}$	$C_8H_{17}COCO_2C_2H_5$	70	$(CH_3-\!\!\bigcirc\!\!-)_3N$	[4]
$\underset{R^2}{\overset{R^1}{>}}C\!\!<\!\!\underset{SR^3}{\overset{SR^3}{}}$ Nu	R^3S-SR^3 $R^1R^2CNu_2$	—	$(Br-\!\!\bigcirc\!\!-)_3N$	[5]

2.9.3 Heteromediatory Systems

In the heteromediatory system, the substrate S is not oxidized by direct electron transfer from S to M_{ox} but by chemical reaction between S and $M_{ox}{}^\cdot$. Many of the mediatory systems which are useful in organic synthesis may be classified into this category. Although some indirect oxidations using heavy metal ions as oxidizing agents have been known as rather classical types of mediatory systems, the most promising systems are those using new mediators which may not easily be prepared by the usual chemical methods. Among a variety of mediators, the redox system consisting of a halide anion and a positive halogen species is one of the most interesting mediators used in organic synthesis.

One of the earliest synthetic reactions in which the halide anion was used as a mediator is the anodic methoxylation of furan in the presence of 0.05 equivalent of ammonium bromide, though the reaction has not been termed a mediated oxidation (3) [6].

$$\text{(furan)} \xrightarrow[\text{NH}_4\text{Br}/\text{CH}_3\text{OH}]{-2e} CH_3O \quad O \quad OCH_3 \qquad Y=73\% \tag{3}$$

In this reaction the participation of the direct oxidation of furan can not be excluded, but the main pathway is the oxidation of bromide anion to bromine followed by its reaction with furan. The formed intermediate is then solvolyzed with methanol to yield the product and to regenerate bromide anion.

A catalytic amount (0.01 equivalent) of potassium iodide has been successfully used in the anodic coupling of active methylene compounds such as ethyl malonate and ethyl acetoacetate (4) [7, 8].

$$
\begin{aligned}
I^- &\xrightarrow{-e} \tfrac{1}{2} I_2 \qquad\qquad K^+ \xrightarrow{+e} K\cdot \\
K\cdot \ + \ &CH_3COCH_2CO_2C_2H_5 \longrightarrow CH_3CO\bar{C}HCO_2C_2H_5 \ + \ K^+ \ + \ \tfrac{1}{2} H_2 \\
CH_3CO\bar{C}HCO_2C_2H_5 \ + \ &\tfrac{1}{2} I_2 \longrightarrow \tfrac{1}{2} \ \underset{CH_3CO\overset{|}{C}HCO_2C_2H_5}{CH_3CO\overset{|}{C}HCO_2C_2H_5} \ + \ I^-
\end{aligned}
\tag{4}
$$

$$Y=40\%$$

If the reaction is performed without potassium iodide very low yields of coupled dimers are obtained. Dimerization of the carbanions of active methylene compounds using iodide as the mediator has been extended to the paired reaction system in which the cathodic reaction is dimerization of ethyl acrylate and diethyl malonate anion is dimerized by the anodic process [9, 10].

After these early investigations, a variety of oxidations using the redox system halide anion/positive halogen species as the mediator have been studied. Some of these oxidation systems are compiled in Table 2.

The use of halogen as the mediator sometimes leads to unique reactions as shown by the example below [25].

$$C_6H_{13}CHO \ + \ (CH_3)_2NH \xrightarrow[\text{KI, t-C}_4\text{H}_9\text{OH, H}_2\text{O}]{-2e} C_5H_{11}COCH_2N(CH_3)_2 \qquad Y=56\% \tag{5}$$

2. Anodic Oxidations

Table 2. Oxidation Using Halide Ion as the Mediator

Substrate	Mediator	Solvent	Product	Yield (%)	Ref.
CH_3CONH_2	NaBr, NaOH	—	$CH_3CONHCONHCH_3$	68	[11]
$C_6H_{13}-CH-CH_3$ I	I_2	CH_3CN	$C_6H_{13}CHCH_3$ $NHCOCH_3$	54	[12]
			$C_5H_{11}CHC_2H_5$ $NHCOCH_3$	22	
1-Iodoadamantane	I_2	CH_3CN	1-Adamantyl-$NHCOCH_3$	60	[12]
CH_3OH, CO	NH_4Br	CH_3OH	$(CH_3O)_2CO$	80	[13]
$(C_2H_5)_2POH$	LiCl	C_2H_5OH	$(C_2H_5O)_2\overset{O}{\overset{\|}{P}}OC_2H_5$	70	[14]
$(CH_2)_{10}\begin{array}{l}-CH-OH\\-CH_2\end{array}$	KI	$t\text{-}C_4H_9OH-C_6H_{14}-H_2O$	$(CH_2)_{10}\begin{array}{l}-C=O\\-CH_2\end{array}$	91	[15]
$C_6H_{13}CH-CH_3$ OH	KI	$t\text{-}C_4H_9OH-H_2O$	$C_6H_{13}\overset{O}{\overset{\|}{C}}CH_3$	99	[15]
$C_8H_{17}OH$	KI	H_2O	$C_8H_{17}O_2CC_7H_{15}$	83	[15]
⬡$-(CH_2)_3OH$	KI	$t\text{-}C_4H_9OH-C_6H_{14}-H_2O$	⬡$-(CH_2)_3O_2C(CH_2)_2-$⬡	84	[15]
$(C_2H_5O)_2POH$ + $(C_2H_5)_2NH$	NaI	CH_3CN	$(C_2H_5O)_2\overset{O}{\overset{\|}{P}}N(C_2H_5)_2$	92	[16]
⬡⬡⬡–OAc	NaBr	$CH_3CN-THF-H_2O$ $-(C_2H_5)_3N$	(epoxide)–OAc	75	[17]
⬡⬡⬡–CO_2CH_3	NaBr	$CH_3CN-THF-H_2O$	(epoxide)–CO_2CH_3	74	[17]
⬡⬡⬡–OH	NaBr	$CH_3CN-H_2O-HCO_2H$	(epoxide)–OH	77	[17]
(phthalimide) + $(\text{cyclohexyl}-S)_2$	NaBr	CH_3CN	(phthalimide)$NS-$cyclohexyl	99	[18]
$(C_2H_5O)_2POH$ + $(C_6H_5S)_2$	NaBr	CH_3CN	$(C_6H_5O)_2\overset{O}{\overset{\|}{P}}S-C_6H_5$	91	[19]

118

Table 2. (continued)

Substrate	Mediator	Solvent	Product	Yield (%)	Ref.
$C_6H_{13}CHCH_3$ OH	poly-(4-vinylpyridine hydrobromide) (PVH)	CH_3CN	$C_6H_{13}COCH_3$	98	[20]
⌬—CHCH₃ OH	PVH	CH_3CN	⌬—COCH₃	91	[20]
$(CH_2)_{10}$ ⟨CHOH / CH₂⟩	PVH	CH_3CN	$(CH_2)_{10}$ ⟨CO / CH₂⟩	82	[20]
$C_8H_{17}OH$	PVH	CH_3CN	$C_7H_{15}CO_2H$	59	[20]
⟨cyclopentanone⟩=O $(C_6H_5Se)_2$	$(C_2H_5)_4NBr$ $MgBr_2$	CH_3CO_2H	⟨cyclopentanone⟩=O SeC_6H_5	97	[21]
CH_3COCH_3 $(C_6H_5Se)_2$	$(C_2H_5)_4NBr$ $MgBr_2$	$CH_3CO_2H-H_2SO_4$	$CH_3COCH_2SeC_6H_5$	88	[21]
⟨cyclohexene⟩ $(C_6H_5Se)_2$	$(C_2H_5)_4NBr$	$CH_3OH-H_2SO_4$	⟨cyclohexane⟩—OCH_3 SeC_6H_5	96	[22]
⟨structure⟩OAc $(C_6H_5Se)_2$	$(C_2H_5)_4NBr$	$CH_3OH-H_2SO_4$	H_3CO⟨structure⟩OAc SeC_6H_5	83	[22]
⟨cyclohexenyl⟩—$OCOCH_3$ $(C_6H_5Se)_2$	$(C_2H_5)_4NBr$	CH_3OH	⟨cyclohexanone⟩=O SeC_6H_5	95	[22]
$C_8H_{17}CONH_2$	KBr	CH_3OH	$C_8H_{17}NHCO_2CH_3$	73	[23]
⟨cyclohexane⟩CH_3 / $CONH_2$	KBr	CH_3OH	⟨cyclohexane⟩CH_3 / NCO	80	[23]
⌬—CH_2OH	KI	$t-C_4H_9OH-H_2O$	⌬—CO_2CH_2—⌬	73	[24]
CH_2O $(CH_3)_2NH$	KI	H_2O	$(CH_3)_2NCHO$	95	[25]
CH_2O ⟨piperidine⟩NH	KI	H_2O	⟨piperidine⟩NCHO	93	[25]

Table 2. (continued)

Substrate	Mediator	Solvent	Product	Yield (%)	Ref.
(structure with CHO group) $(C_4H_9)_2NH$	KI	$t-C_4H_9OH-H_2O$	(structure with $N(C_4H_9)_2$)	60	[25]
$C_6H_5CH_2CHCO_2CH_3$ $\vert$ $NHCO_2CH_3$	NaCl	CH_3OH	OCH_3 $\vert$ $C_6H_5CH_2CCO_2CH_3$ $\vert$ $NHCO_2CH_3$	90	[26]

This rather strange reaction may be explained by the following reaction mechanism:

$$I^- \xrightarrow{-2e} I^+$$

$$RCH_2CHO + HNR'_2 \longrightarrow RCH=CHNR'_2 \xrightarrow{I^+} RCH\cdots CHNR' \xrightarrow{OH^-}$$

$$\underset{\overset{\vert}{I}\ \overset{\vert}{OH}}{RCHCHNR'_2}\ (or\ \underset{\overset{\vert}{OH}\ \overset{\vert}{I}}{RCH-CHNR'_2}) \xrightarrow{base} R-CH-CHNR'_2 \longrightarrow RCCH_2NR'_2 \tag{6}$$

The electrochemical oxidation using a mediator sometimes gives an entirely different result as compared with the direct oxidation. The direct oxidation of the N,N'-biscarbamate of lysine methyl ester in methanol leads to methoxylation of the carbon atom adjacent to the ω-amino group. On the other hand, the oxidation of the same compound in methanol containing sodium chloride results in methoxylation at the position α to the methoxy-carbonyl group (7) [26]. This difference may be explained as follows.

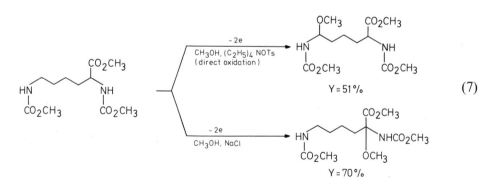

$$\tag{7}$$

The oxidation potential required for the removal of one electron from the lone-pair electrons of the nitrogen of the ω-amino group is lower than that required for the removal of an electron from the α-amino group. Hence, the direct methoxylation takes place at the neighboring carbon of the ω-amino group. On the other hand, if the reaction is carried out in the presence of sodium chloride, the chloride ion and its oxidized species act as a mediator. The active species of the mediator reacts with the α-amino group to yield an N-chloro derivative which is converted into an imino compound by reaction with a base formed in situ. The addition of methanol to the imino group gives the α-methoxylated product.

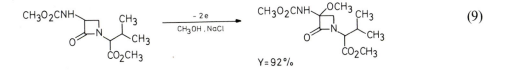

(8)

This α-methoxylation is very useful for the methoxylation of some β-lactams.

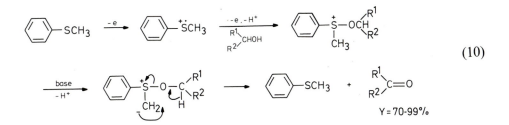

(9)

Besides halide ions organic sulfides are also efficient mediators. The oxidation of secondary alcohols to ketones has been successfully achieved by using methyl phenyl sulfide as the mediator (10) [27].

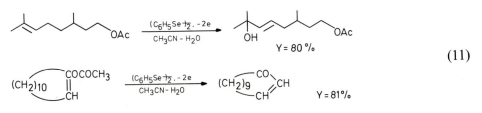

(10)

It is remarkable that carbon—carbon double bonds are completely inert in these oxidations.

In the following reactions C_6H_5SeOH is used as a mediator (11) [28].

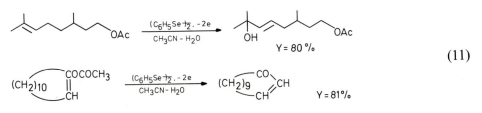

(11)

121

2. Anodic Oxidations

In the oxidation of 2-butanol lithium or tetraethylammonium nitrate may be used as a mediator (12) [29].

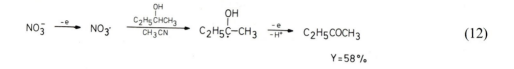

$$NO_3^- \xrightarrow{-e} NO_3^{\cdot} \xrightarrow[CH_3CN]{C_2H_5\overset{\overset{\displaystyle OH}{|}}{C}HCH_3} C_2H_5\overset{\overset{\displaystyle OH}{|}}{C}-CH_3 \xrightarrow[-H^+]{-e} C_2H_5COCH_3 \qquad (12)$$

Y = 58 %

2.9.4 Double Mediatory System

As described above, the mediatory system is an effective tool to oxidize the substrates that can not be readily oxidized by the direct method. Further development of this concept has led to the combination of two types of mediators (Fig. 2). As a result, the oxidation of substrates is achieved at a potential which is far lower than that required when the system contains only one type of meadiator [30].

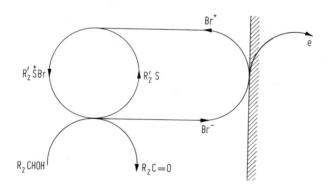

Fig. 2. Double Mediatory System

In this system, the potential (E_p = 1.1 V vs. SCE) of the oxidation of Br^- to Br^+ is the lowest, and the oxidation of $R_2'S$ to $R_2'S^{+\cdot}$ does not take place at this potential. As described above, alcohols such as R^1R^2CHOH are oxidized by $R_2'S^{+\cdot}$ whereas Br^+ itself is not sufficiently reactive as to oxidize alcohols to ketones in satisfactory yields. When both mediators are combined as depicted in Fig. 2, however, the oxidation of alcohols may be achieved at a considerably lower potential than that necessary for the oxidation of $R_2'S$ to $R_2'S^{+\cdot}$. The reaction sequence is shown in scheme (13), methyl octyl sulfide (E_p = 1.93 V vs. SCE) being used as the mediator.

122

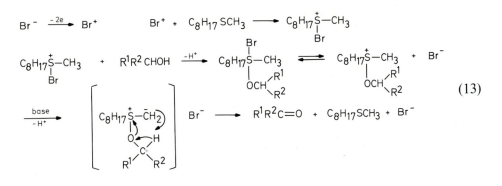

$$(13)$$

The yields of the obtained ketones are in the range of 80–94 %. The mediatory system shown in Fig. 2 can be called a double mediatory system.

References

1. Schmidt, W., Steckhan, E.: J. Electroanal. Chem. *89*, 215 (1978)
2. Schmidt, W., Steckhan, E.: Angew. Chem. Int. Ed. Engl. *17*, 673 (1978)
3. Schmidt, W., Steckhan, E.: J. Electroanal. Chem. *101*, 123 (1979)
4. a) Platem, M., Steckha, E.: Tetrahedron Lett. *21*, 511 (1980)
   b) See also Schmidt, W., Steckhan, E.: Chem. Ber. *113*, 577 (1980)
5. Martigny, P., Simonet, J.: J. Electroanal. Chem. *111*, 133 (1980)
6. Clauson-Kaas, N., Limborg, F., Glens, K.: Acta Chem. Scand. *6*, 531 (1952)
7. Okubo, T., Tsutsumi, S.: Bull. Chem. Soc. Jpn. *37*, 1794 (1964)
8. Jansson, R. E. W., Tomov, N. R.: Electrochim. Acta *25*, 497 (1980)
9. Thomas, H. G., Lux, E.: Tetrahedron Lett. *1972*, 965
10. Baizer, M. M., Hallcher, R. C.: J. Electrochem. Soc. *123*, 809 (1976)
11. Lanchen, M.: J. Chem. Soc. *1950*, 748
12. Miller, L. L., Watkins, B. F.: Tetrahedron Lett. *1974*, 4495
13. Cipris, D., Mador, I. L.: J. Electrochem. Soc. *125*, 1954 (1978)
14. Ohmori, H., Nakai, S., Sekiguchi, M., Masui, M.: Chem. Pharm. Bull. *27*, 1700 (1979)
15. Shono, T., Matsumura, Y., Hayashi, J., Mizoguchi, M.: Tetrahedron Lett. *1979*, 165
16. Torii, S., Sayo, N., Tanaka, H.: Tetrahedron Lett. *1979*, 4471
17. a) Torii, S., Uneyama, K., Ono, M., Tazawa, H., Matsunami, S.: Tetrahedron Lett. *1979*, 4661
    b) See also Torii, S., Uneyama, K., Matsumori, S.: J. Org. Chem. *45*, 16 (1980)
18. Torii, S., Tanaka, H., Ukida, M.: J. Org. Chem. *44*, 1554 (1979)
19. Torii, S., Tanaka, H., Sayo, N.: J. Org. Chem. *44*, 2938 (1979)
20. Yoshida, J., Nakai, R., Kawabata, N.: J. Org. Chem. *45*, 5269 (1980)
21. Torii, S., Uneyama, K., Handa, K.: Tetrahedron Lett. *21*, 1863 (1980)
22. a) Torii, S., Uneyama, K., Ono, M.: Tetrahedron Lett. *21*, 2741 (1980)
    b) See also Torii, S., Uneyama, K., Ono, M.: Tetrahedron Lett. *21*, 2653 (1980)
23. Shono, T., Matsumura, Y., Yamane, S., Kashimura, S.: Chem. Lett. *1982*, 565

24. Dixit, G., Rastogi, R., Zutshi, K.: Electrochim. Acta 27, 561 (1982)
25. Shono, T., Matsumura, Y., Hayashi, J., Usui, M., Yamane, S.: Acta Chem. Scand., B37, 491 (1983)
26. Shono, T., Matsumura, Y., Inoue, K.: J. Org. Chem. 48, 1388 (1983)
27. Shono, T., Matsumura, Y., Mizoguchi, M., Hayashi, J.: Tetrahedron Lett. 1979, 3861
28. Torii, S., Uneyama, K., Ono, M., Bannou, T.: J. Am. Chem. Soc. 103, 4606 (1981)
29. Leonard, J. E., Scholl, P. C., Steckel, T. P., Lentsch, S. L., Van. De Mark, M. R.: Tetrahedron Lett. 21, 4695 (1980)
30. Shono, T., Matsumura, Y., Hayashi, J., Mizoguchi, M.: Tetrahedron Lett. 21, 1867 (1980)

3. Cathodic Reductions

In contrast to the anodic oxidation, the cathodic reduction is not classified by the structure of the starting compounds but mainly by the reaction patterns involved, though the actual reaction mechanism has in many cases not been elucidated so far.

3.1 Cathodic Addition, Substitution, and Coupling

Since in electroorganic reactions the definition of addition, coupling, and substitution cannot clearly be made, these processes are characterized as follows.

A process in which cathodically generated radicals or anions add to unsaturated bonds including carbonyl groups is called addition, coupling is defined as a combination occurring between radical species including anion radicals, and substitution as a process which is similar to nucleophilic substitution (S_N2).

3.1.1 Addition

Cathodically generated heteroatomic anions may add to acylating reagents to yield the corresponding acylated products as exemplified below (1–4).

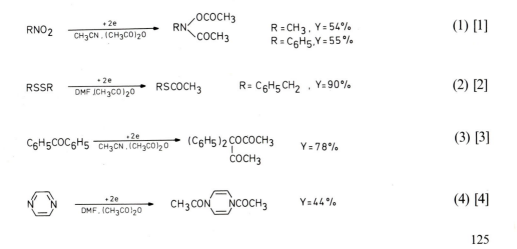

3. Cathodic Reductions

The synthesis of carbonyl compounds by acylation of carbanions formed by cathodic reduction of olefins or benzyl halides is undoubtedly one of the most useful methods of synthesizing ketones. Some typical reactions are summarized in Table 1.

The electroreductively generated carbanions may also add to the carbonyl groups of aldehydes and ketones.

Table 1. Formation of Carbonyl Compounds by Reductive Acylation

Substrate	Acylating agent	Solvent/Supporting electrolyte	Product	Yield %	Ref.
(anthracene structure)	$(CH_3CO)_2O$	DMF/TBAI[a]	(dihydroanthracene with H_3C $OCOCH_3$)	66–75	[5]
$CH_2=CHCO_2CH_3$	$(CH_3CO)_2O$	CH_3CN/TEAT[b]	$CH_3COCH_2CH_2CO_2CH_3$	62	[6]
$CH_3CH=CHCO_2CH_3$	$(C_3H_7CO)_2O$	CH_3CN/TEAT[b]	$C_3H_7COCHCH_2CO_2CH_3$ CH_3	82	[6]
$C_6H_5CH=CHCO_2C_2H_5$	$(C_2H_5CO)_2O$	CH_3CN/TEAT[b]	$C_2H_5COCHCH_2CO_2C_2H_5$ C_6H_5	75	[6] [7]
$C_6H_5CH=CHCN$	$(CH_3CO)_2O$	CH_3CN/TEAT[b]	$CH_3COCHCH_2CN$ C_6H_5	76	[6] [7]
$C_6H_5CH_2Cl$	C_3H_7COCl	CH_3CN/TEAT[b]	$C_6H_5CH_2COC_3H_7$	69	[8]
C_6H_5CHCl CH_3	C_3H_7COCl	CH_3CN/TEAT[b]	$C_6H_5CHCOC_3H_7$ CH_3	71	[8]
$C_6H_5C=CH_2$ CH_3	CH_3CN	CH_3CN/LiClO$_4$	$C_6H_5CHCH_2COCH_3$ CH_3	68	[9]
$C_6H_5CH=CH_2$	DMF	DMF/LiClO$_4$	$C_6H_5CHCH_2CHO$ CHO	82	[9]
$C_6H_5CH=CH-$(pyridyl, N=)	$(CH_3CO)_2O$	DMF/TBAI[a]	$C_6H_5CHCH_2-$(pyridyl, N=) $COCH_3$	45	[10]
$C_6H_5C=CHCO_2C_2H_5$ $OCOCH_3$	$(CH_3CO)_2O$	DMF/TBAI[a]	$COCH_3$ $C_6H_5CHCH_2CO_2C_2H_5$ $OCOCH_3$	68	[10]

[a] Tetrabutylammonium iodide, [b] Tetraethylammonium *p*-toluenesulfonate.

126

Trichloromethyl carbanion generated by cathodic reduction of carbon tetrachloride in DMF adds to aldehydes yielding trichloromethylcarbinols (5) [11].

$$CCl_4 \xrightarrow{+2e} \bar{C}Cl_3 \xrightarrow{C_6H_5CHO} \underset{\underset{OH}{|}}{C_6H_5CHCCl_3} \qquad Y = 70\% \qquad (5)$$

When this reaction is carried out in the presence of a sufficient amount of chloroform, it proceeds as a chain reaction (6) [12].

$$CCl_4 \xrightarrow{+2e} \bar{C}Cl_3 + Cl^-, \quad \bar{C}Cl_3 + RCHO \longrightarrow \underset{\underset{O^-}{|}}{R-CHCCl_3}$$

$$\underset{\underset{O^-}{|}}{R-CHCCl_3} + CHCl_3 \longrightarrow \underset{\underset{OH}{|}}{R-CHCCl_3} + \bar{C}Cl_3$$

$R = CH_3O-\bigcirc- \qquad Y = 90\%$

$R = \qquad Y = 81\%$

$R = \underset{CH_3}{\overset{CH_3}{\diagup}}\diagdown \qquad Y = 90\%$

(6)

The highest current efficiency observed is about $3 \times 10^3 \%$ and the chemical yields are often around 90%. The obtained trichloromethylcarbinols are highly promising starting materials in organic synthesis as shown by the following diagram [13].

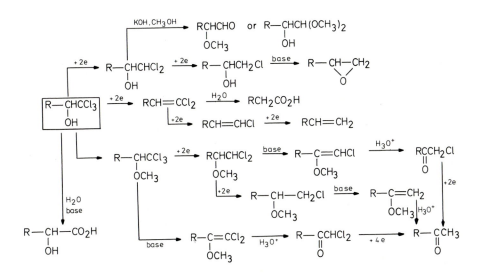

The formation of α-hydroxy aldehydes from trichloromethylcarbinols is successfully applied to the synthesis of erythrose, threose and erythrulose from glyceraldehyde (7), (8) [14].

127

3. Cathodic Reductions

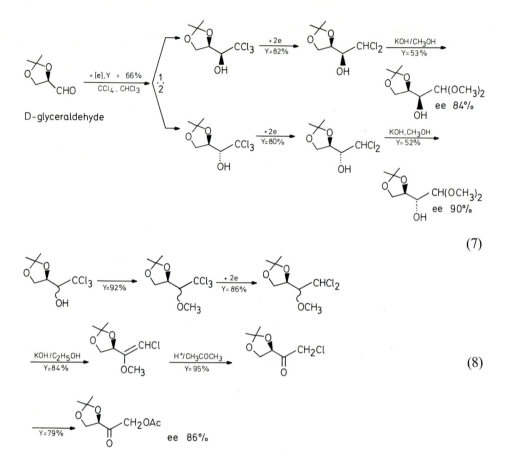

(7)

(8)

The carbanion formed by the reduction of ethyl trichloroacetate also adds to carbonyl groups (9) [11].

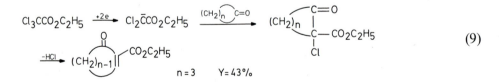

(9)

The reaction sequence (9) can be carried out as a chain reaction by using dichloroacetic acid ester together with trichloroacetate (10) [15].

$$Cl_3CCO_2CH_3 \xrightarrow{+2e} Cl_2\bar{C}CO_2CH_3 \xrightarrow{RCHO} R-CHCCl_2CO_2CH_3 \xrightarrow{CHCl_2CO_2CH_3} RCHCCl_2CO_2CH_3$$

(10)

A carbanion generated from diethyl trichloromethylphosphate or triethyl phosphonodichloroacetate can be used in the Wittig-Horner reaction (11) (12) [16].

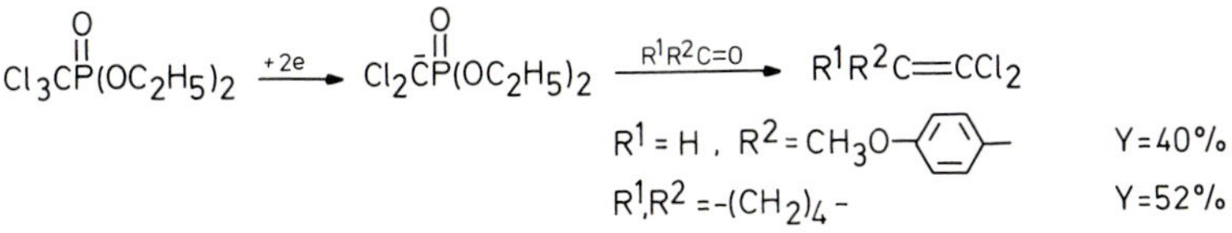

$$Cl_3C\overset{O}{\overset{||}{P}}(OC_2H_5)_2 \xrightarrow{+2e} Cl_2\overset{O}{\overset{||}{\bar{C}}}P(OC_2H_5)_2 \xrightarrow{R^1R^2C=O} R^1R^2C=CCl_2$$

$$R^1 = H, \quad R^2 = CH_3O-\!\!\!\left\langle \right\rangle\!\!\!- \qquad Y = 40\%$$
$$R^1,R^2 = -(CH_2)_4- \qquad Y = 52\%$$

(11)

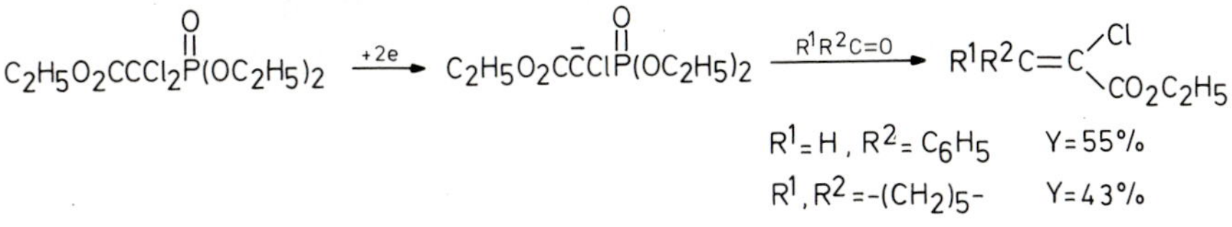

$$C_2H_5O_2CCCl_2\overset{O}{\overset{||}{P}}(OC_2H_5)_2 \xrightarrow{+2e} C_2H_5O_2C\overset{O}{\overset{||}{\bar{C}}}ClP(OC_2H_5)_2 \xrightarrow{R^1R^2C=O} R^1R^2C=C\overset{Cl}{\underset{CO_2C_2H_5}{\diagup}}$$

$$R^1 = H, \quad R^2 = C_6H_5 \qquad Y = 55\%$$
$$R^1, R^2 = -(CH_2)_5- \qquad Y = 43\%$$

(12)

The addition of carbanions formed from activated olefins to carbonyl compounds in sulfuric acid has no synthetic value, due to the low material yield and the lack of generality. However, if the same reaction is carried out in a neutral solvent in the presence of trimethylchlorosilane satisfactory yields of lactones with respect to a variety of activated olefins and carbonyl compounds are obtained (13) [17].

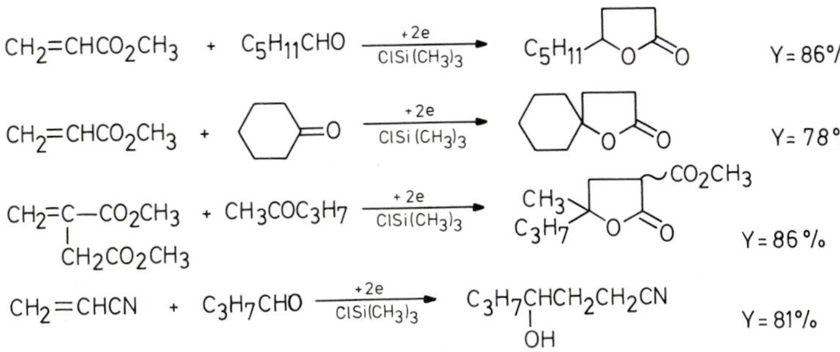

Lactones are also formed by the addition of carbanions, generated by reductive cleavage of carbon-heteroatom bonds, to aldehydes (14) [18].

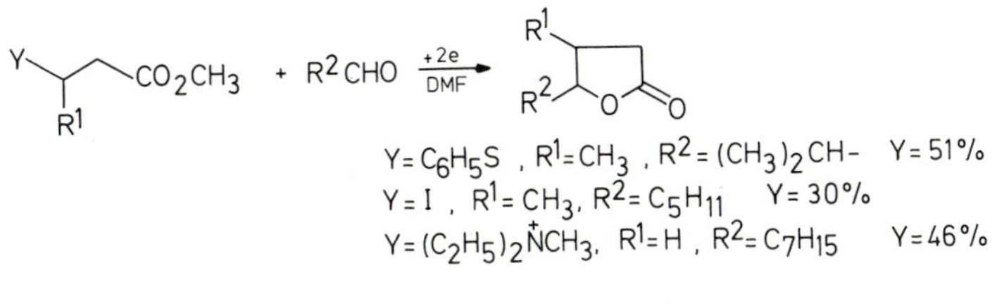

$$Y = C_6H_5, \quad R^1 = CH_3, \quad R^2 = (CH_3)_2CH- \quad Y = 51\%$$
$$Y = I, \quad R^1 = CH_3, \quad R^2 = C_5H_{11} \quad Y = 30\%$$
$$Y = (C_2H_5)_2\overset{+}{N}CH_3, \quad R^1 = H, \quad R^2 = C_7H_{15} \quad Y = 46\%$$

(14)

3. Cathodic Reductions

Similar carbanions also add to carboxylic acid anhydrides to yield γ-keto esters (15) [18].

$$
\text{CH}_3\text{—N}^{+}\overset{}{\underset{R^1}{\big|}}\text{—} \text{CO}_2\text{CH}_3 \; \text{I}^{-} \quad + \quad (R^2CO)_2O \xrightarrow[\text{DMF}]{+2e} R^2\overset{O}{\overset{\|}{C}}\underset{R^1}{\big|}\text{CO}_2\text{CH}_3
$$

(15)

$$
\begin{aligned}
&R^1 = CH_3,\ R^2 = C_3H_7 \qquad\quad Y = 70\% \\
&R^1 = CH_3,\ R^2 = (CH_3)_2CH- \quad Y = 71\%
\end{aligned}
$$

The reductive alkylation of carbonyl compounds with simple quaternary ammonium salts has only been observed in a few cases in which the actual reaction route is not the addition of carbanions formed from the ammonium salts to carbonyl compounds, but the reduction of carbonyl compounds to anionic intermediates followed by alkylation of these intermediates by the ammonium salts (16) [19, 20].

$$
(C_2H_5)_4N\text{OTs} \; + \; R^1COR^2 \xrightarrow[\text{DMF}]{+2e} C_2H_5\overset{R^2}{\underset{R^1}{\big|}}COH
$$

(16)

$$
\begin{aligned}
&R^1 = CH_3,\ R^2 = C_5H_{11} \qquad\qquad\ Y = 55\% \\
&R^1 = CH_3,\ R^2 = CH_2\!=\!CHCH_2CH_2 \quad Y = 65\% \\
&R^1, R^2 = -(CH_2)_5- \qquad\qquad\qquad Y = 51\%
\end{aligned}
$$

The formation of a furan derivative has been observed in the reduction of phenacyl bromide (17) [21]. This reaction seems to involve the addition of a phenacyl anion to the starting compound.

$$
2\,C_6H_5COCH_2Br \xrightarrow{+2e} \quad \text{(furan derivative)} \qquad Y = 80\%
$$

(17)

The trapping of cathodically generated anionic species by carbon dioxide (carboxylation) has been studied extensively. Some typical results are shown in Table 2.

The intermolecular addition of anionic species generated by cathodic reduction of ketones to nonactivated olefines is only achieved in low yields (18) [29] whereas the intramolecular addition, i.e. reductive cyclization of nonconjugated olefinic ketones, proceeds in high yields with excellent regio- and stereo-selectivities (19). Some of typical cycloadducts are compiled in Table 3.

$$
R^1COR^2 \; + \; R^3CH\!=\!CH_2 \xrightarrow[\text{isoPrOH -dioxane}]{\cdot 2e} R^2\overset{R^1}{\underset{OH}{\big|}}C\text{—}CH_2CH_2R^3
$$

(18)

$$
\begin{aligned}
&R^1 = CH_3,\ R^2 = C_2H_5,\ R^3 = C_6H_{13} \qquad\ Y = 22\% \\
&R^1 = CH_3,\ R^2 = C_6H_{13},\ R^3 = C_6H_{13} \quad Y = 20\%
\end{aligned}
$$

Table 2. Electrocarboxylation [28]

Substrate	Solvent	Supporting electrolyte	Product	Yield %	Ref.
$C_6H_5CH=NC_6H_5$	—	$(C_2H_5)_4NOTs$	$C_6H_5CH-NHC_6H_5$ $\qquad\quad \mid$ $\qquad\quad CO_2H$	60	[22]
$CH_2=CHCO_2CH_3$	CH_3CN	$(C_2H_5)_4NOTs$	$CH_3O_2CCH_2CH(CO_2CH_3)_2$	61	[23]
$CH_2=CHCN$	CH_3CN	$(C_2H_5)_4NOTs$	$CH_3O_2CCHCH_2CO_2CH_3$ $\qquad\qquad \mid$ $\qquad\qquad CN$	41	[23]
$CH_2=CHCOCH_3$	CH_3CN	$(C_2H_5)_4NOTs$	$\{ CH_3COCH_2CH_2CO_2CH_3$ $\{ CH_3O_2CCHCH_2CO_2CH_3$ $\qquad\qquad \mid$ $\qquad\qquad COCH_3$	22 16	[23]
$\begin{array}{c} CO_2CH_3 \\ \diagup \\ \diagdown \\ CO_2CH_3 \end{array}$	CH_3CN	$(C_2H_5)_4NOTs$	$(CH_3O_2C)_2CHCHCHCH(CO_2CH_3)_2$ $\qquad\qquad\quad \mid \quad\mid$ $\qquad\qquad\quad CH_3O_2C \;\; CO_2CH_3$	46	[24]
$(CH_2)_2 \begin{array}{c} CH=CHCO_2CH_3 \\ \diagup \\ \diagdown \\ CH=CHCO_2CH_3 \end{array}$	CH_3CN	$(C_2H_5)_4NOTs$	cyclopentane with CO_2CH_3, CO_2CH_3, $CH(CO_2CH_3)_2$	72	[24]
$(CH_2)_3 \begin{array}{c} CH=CHCO_2CH_3 \\ \diagup \\ \diagdown \\ CH=CHCO_2CH_3 \end{array}$	CH_3CN	$(C_2H_5)_4NOTs$	cyclohexane with $CH(CO_2CH_3)_2$, $CH(CO_2CH_3)_2$	50	[24]
$C_6H_5CHCl_2$	DMF	$(C_4H_9)_4NBr$	$C_6H_5CHCO_2H$ $\qquad\quad \mid$ $\qquad\quad OH$	9.8	[25]
chloroquinoline	DMF	$(C_4H_9)_4NI$	chloroquinoline with CO_2CH_3 (4-position) chloroquinoline with CO_2CH_3 (2-position)	50 5	[26]
$\begin{array}{c} C_6H_5 \\ \diagdown \\ \diagup \\ C_6H_5 \end{array} C=NC_6H_5$	CH_3CN	$(C_2H_5)_4NClO_4$	$\begin{array}{c} C_6H_5 \\ \diagdown \\ \diagup \\ C_6H_5 \end{array} C-NHC_6H_5$ $\qquad\qquad \mid$ $\qquad\qquad CO_2H$	87	[27]

$$R^1CH=CH(CH_2)_n COR^2 \xrightarrow[CH_3OH-dioxane]{+2e} R^2C \overset{(CH_2)_n}{\underset{OH}{\diagup \diagdown}} \underset{}{CHCH_2R^1} \qquad (19)$$

When vinyl acetate is used as the unsaturated compound, the intermolecular addition of the anionic species, generated from aromatic ketones, takes place to the carbon—carbon double bond smoothly (20) [33].

3. Cathodic Reductions

Table 3. Intramolecular Cycloaddition of Nonconjugated Unsaturated Ketones

Ketone	Solvent	Product	Yield (%)	Ref.
$CH_3CO(CH_2)_3CH=CH_2$	CH_3OH–dioxane		98	[30] [31]
$CH_3CO(CH_2)_3CH=CH-CH_3$	CH_3OH–dioxane		86	[31]
$CH_3CO(CH_2)_3CH=C(CH_3)_2$	DMF		54	[31]
$CH_3CO(CH_2)_4CH=CH_2$	DMF		75	[31]
$CH_3CO(CH_2)_2NCH_2CH=CH_2$ $|$ C_3H_7	DMF		34	[31]
	CH_3OH–dioxane		87	[31]
	CH_3OH–dioxane		65	[31]
	CH_3OH–dioxane		69	[31]
$CH_3CO(CH_2)_3C\equiv CH$	DMF		94	[32]
	DMF		75	[32]

$$ArCOAr' \; + \; CH_2=CHOAc \xrightarrow[\text{DMF}]{+2e} \underset{\substack{| \\ OH}}{\overset{\substack{Ar \\ \diagdown \\ Ar' \diagup}}{C}} - \overset{\substack{OAc \\ |}}{C}HCH_3 \qquad Ar, Ar' = C_6H_5 \quad Y = 67\% \qquad (20)$$

Although the radicals and anions generated by the cathodic reduction of alkyl halides also add to unsaturated bonds, the reduction of systems consisting of alkyl halides and olefins does not always lead to such an addition reaction since olefins are often reduced faster than alkyl halides [34, 35]. The reaction of alkyl halides with anionic species formed by the

132

cathodic reduction of unsaturated compounds will be referred to substitution.

The reduction of a solution of carbon tetrachloride in chloroform easily yields trichloromethyl anions which add to active olefins like acrylonitrile with high current efficiency (347%) (21) [36].

$$CCl_4 \xrightarrow{+2e} \bar{C}Cl_3 + Cl^-$$
$$CH_2=CHCN + \bar{C}Cl_3 \longrightarrow CCl_3CH_2\bar{C}HCN \tag{21}$$
$$CCl_3CH_2\bar{C}HCN + CHCl_3 \longrightarrow CCl_3CH_2CH_2CN + \bar{C}Cl_3$$

The anion formed from ethyl trichloroacetate also adds to olefins in a similar way.

The reduction of 1-bromo-5-decyne in DMF at -2.70 V $vs.$ SCE yields 1-butylcyclohexene (38%) as a peculiar product (22) [37].

$$C_4H_9C{\equiv}C(CH_2)_4Br \xrightarrow[DMF]{+2e} \bigcirc\!\!\!\!-C_4H_9 \tag{22}$$

The anion formed by the cathodic reduction of ethyl N-chlorocarbamate adds to active olefins to yield the corresponding amine derivatives (23) [38].

$$C_2H_5O_2CNHCl + CH_2=CHCO_2CH_3 \xrightarrow{+2e}_{CH_3OH-CHCl_3} C_2H_5O_2CNHCH_2CH_2CO_2CH_3 \tag{23}$$
$$Y = 61\%$$

3.1.2 Substitution

The S_N-2 type substitution may take place when anions are cathodically generated in the presence of alkyl halides and similar reagents. The reduction of a solution of an α,β-unsaturated compound and alkyl halide in hexamethylphosphoric triamide (HMPA) yields the corresponding alkylated compound as exemplified by Eq. (24) [39].

$$C_2H_5O_2CCH=CHCO_2C_2H_5 + C_2H_5Br \xrightarrow[HMPA]{+2e} C_2H_5-CHCO_2C_2H_5 \tag{24}$$
$$\underset{CH_2CO_2C_2H_5}{|} \quad Y = 45\%$$

This alkylation is not a Michael addition of ethyl anion to diethyl maleate but a substitution of ethyl bromide by the anionic species formed by the reduction of the maleate. This reductive alkylation is often accompanied by α,β-dialkylation. The alkylation of α,β-unsaturated compounds may lead to intramolecular cyclization as shown in Table 4.

The electroreductive cyclization of ω-bromoalkylidenemalonates (Table 4) is a unique reaction since the cyclization of the same compounds caused by a hydride transfer agent gives different products (25).

3. Cathodic Reductions

Table 4. Reductive Cyclization

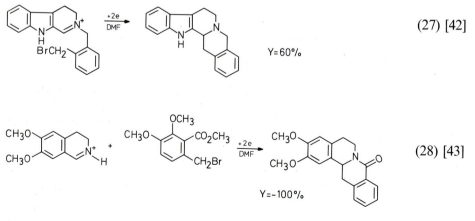

Substrate	Product	Yield (%)	Ref.
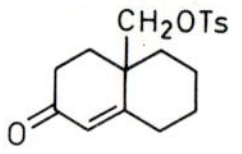	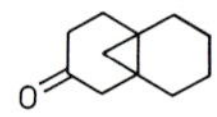	98	[40]
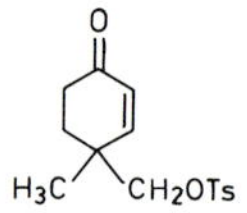	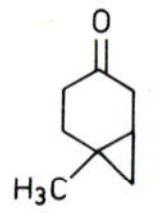	80	[40]
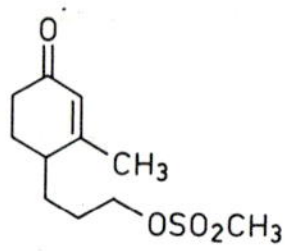	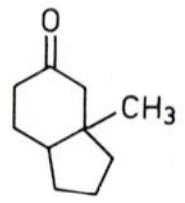	82	[40]
$Br(CH_2)_3CH{=}C(CO_2CH_3)_2$	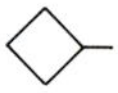 $-CH(CO_2CH_3)_2$	65	[41]
$Br(CH_2)_4CH{=}C(CO_2CH_3)_2$	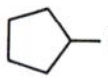 $-CH(CO_2CH_3)_2$	60	[41]

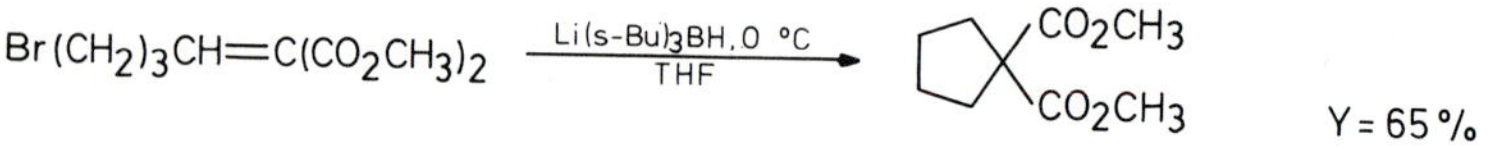

$$Br(CH_2)_3CH{=}C(CO_2CH_3)_2 \xrightarrow[\text{THF}]{\text{Li(s-Bu)}_3BH, 0\ ^\circ C} \quad \begin{array}{c} CO_2CH_3 \\ CO_2CH_3 \end{array} \qquad Y=65\% \tag{25}$$

The similar alkylation of anions formed by the electroreduction of iminium cations has been shown to be highly useful in the synthesis of alkaloid-type compounds (26).

$$\begin{array}{c} R^1 \\ R^2 \end{array} C{=}\overset{+}{N} \begin{array}{c} R^3 \\ R^4 \end{array} \xrightarrow{+2e} R^1R^2\bar{C}{-}NR^3R^4 \xrightarrow{R^5X} R^1R^2R^5CNR^3R^4 \tag{26}$$

Some typical reactions are shown below.

$$\xrightarrow[\text{DMF}]{+2e} \qquad Y=60\% \tag{27} [42]$$

$$\xrightarrow[\text{DMF}]{+2e} \qquad Y={\sim}100\% \tag{28} [43]$$

134

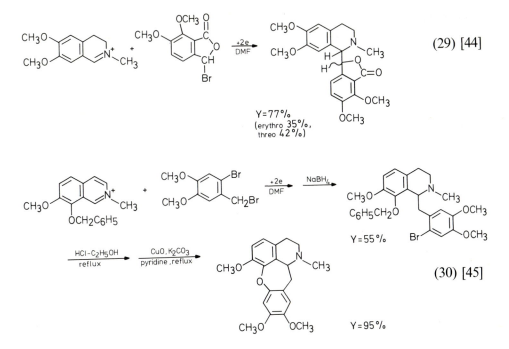

Y=77%
(erythro 35%,
threo 42%)

(29) [44]

Y=55%

CuO, K₂CO₃
pyridine, reflux

HCl-C₂H₅OH
reflux

(30) [45]

Y=95%

Azobenzene and anils are also alkylated by the electroreductive method.

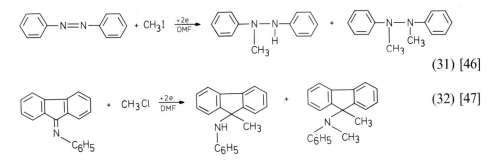

(31) [46]

(32) [47]

The ratio of mono- and di-alkylation can be controlled by modification of the reaction conditions.

When unsaturated compounds are reduced in the presence of α,ω-dihaloalkanes, α,ω-alkanedioyl dichlorides or ω-halogeno acid chlorides, cyclized products are obtained in moderate yields [48].

(33)

Y=59%

(34)

Y=81%

3. Cathodic Reductions

Table 5. Alkylation of Cathodically Generated Anions

Substrate	Alkylhalide	Product	Yield (%)	Ref.
(pyrene)	t-C_4H_9Cl	(pyrene)X	52	[49]
$C_6H_5COC_6H_5$	t-C_4H_9Cl	$(C_6H_5)_2C$-Bu-t, OH	60	[50]
CH_3O-⟨⟩-CO-⟨⟩-OCH_3	t-C_4H_9Cl	$(CH_3O$-⟨⟩-$)_2C$-Bu-t, OH	72	[50]
$\begin{array}{c}CH_2CO\\ \qquad NH\\ CH_2CO\end{array}$	C_4H_9Br	$\begin{array}{c}CH_2CO\\ \qquad N-C_4H_9\\ CH_2CO\end{array}$	90	[51]
$\begin{array}{c}CH_2CO\\ \qquad NBr\\ CH_2CO\end{array}$	CH_3OTs	$\begin{array}{c}CH_2CO\\ \qquad NCH_3\\ CH_2CO\end{array}$	75	[52]
$(C_6H_5S\rightarrow)_2$	CH_3Cl	$C_6H_5SCH_3$	83.5	[53]
$(CH_3S\rightarrow)_2$	$C_6H_5CH_2Cl$	$CH_3SCH_2C_6H_5$	63	[53]
⟨⟩-NO_2	C_4H_9I	⟨⟩-NOC_4H_9, C_4H_9	83	[54]
(tetramethyl dibromo ketone)	CH_3OH (solvent)	(cyclopropane: HO, OCH$_3$, H$_3$C, CH$_3$, H$_3$C, CH$_3$)	~100	[55]

Further examples of the substitution of alkyl halides by cathodically generated anions are summarized in Table 5.

Electroreductive substitution may be applied to aromatic nuclear substitution, though the active intermediates are not anions but radicals.

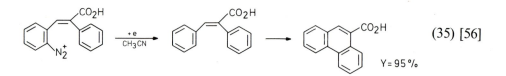

$$(35)\ [56]$$

$$Y = 95\%$$

Further examples are listed in Table 6.

Table 6. Electroreductive Aromatic Substitution

Starting material	Product	Yield (%)	Ref.
		86	[57]
		46	[58]
		59	[59]
		69	[60]

3.1.3 Coupling

The formation of pinacols by the electroreduction of aldehydes or ketones is a typical example of the coupling of radicals or anion radicals generated by cathodic reduction and has been extensively studied. Since the basic pattern of the electroreductive pinacol coupling has already been reviewed [61], only some of the couplings which are useful in organic synthesis are described.

The cathodic pinacol coupling is one of the most efficient methods for the preparation of cyclopropanediols (36) [62].

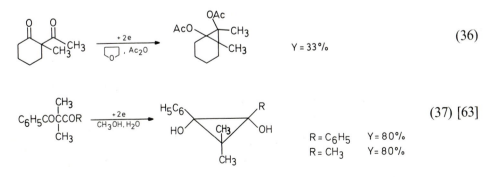

$$Y = 33\%$$ (36)

$$R = C_6H_5 \quad Y = 80\%$$
$$R = CH_3 \quad Y = 80\%$$ (37) [63]

137

3. Cathodic Reductions

The pinacolization of α,β-unsaturated carbonyl compounds seems to be a useful reaction in organic synthesis, although often a mixture of hydro-dimers coupled at the β-carbon is mainly obtained in the cathodic reduction. If the β-position is blocked or sterically hindered, pinacolization is achieved in a satisfactory yield.

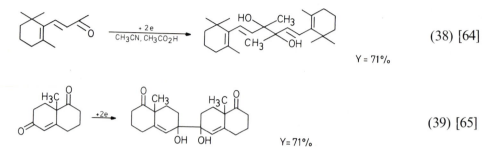

(38) [64]

Y = 71%

(39) [65]

Y= 71%

The electroreduction of a highly conjugated aldehyde, retinal, does not give a pinacol in high yield under the usual reaction conditions but, when diethyl malonate is used as a proton donor, the pinacol of retinal is obtained in 50% yield [66].

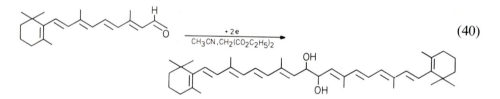

(40)

Stereochemistry is one of the most interesting problems in the electro-reductive pinacolization and has been studied extensively [67–71], although it cannot always be controlled effectively.

Analogously to carbonyl compounds, imines also dimerize under conditions similar to those applied in pinacolization [72].

$$RCH=NR' \xrightarrow[\text{C}_2\text{H}_5\text{OH}-\text{CH}_3\text{CO}_2\text{CH}_3-\text{H}_2\text{O}]{+2e} \begin{matrix} R-CH-NR' \\ R-CH-NR' \end{matrix}$$

$R,R'= CH_3-\langle\ \rangle-$ Y= 73%

$R=Cl-\langle\ \rangle-$, $R'=\langle\ \rangle-$ Y= 88%

(41)

The electroreductive hydrodimerization of α,β-unsaturated compounds is one of the most successful reactions utilized in organic synthesis, the dimerization of acrylonitrile to adiponitrile representing a typical reaction [73]. The mechanism of the electrohydrodimerization of activated olefins has been studied in detail and found to be coupling of two anion radicals, although anion radical-substrate coupling predominates in some cases [74].

138

Anion radical coupling $\quad 2\,R^{-\cdot} \longrightarrow R^{-}\!-\!R^{-}$ (42)

Anion radical-substrate coupling
$$R^{-\cdot} + R \longrightarrow R^{-}\!-\!R\cdot$$
$$R^{-}\!-\!R\cdot + R^{-\cdot} \longrightarrow R^{-}\!-\!R^{-} + R$$ (43)

Since the electrohydrodimerization of typical activated olefins has already been summarized in detail [61b, 75], only few examples applied to organic synthesis are discussed in this section.

A perhydrophenanthrene ring system has been synthesized stereospecifically through hydrodimerization [76].

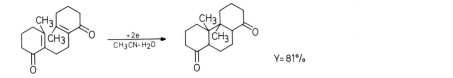

CH3
CH3
O
O
+2e
CH3CN-H2O
CH3
CH3
O
O
Y= 81%

(44)

The reduction of benzene-1,2-diacrylate yields a peculiar dimer in 29% yield [77].

CO2CH3
CO2CH3
+2e
DMF-H2O
CH2CO2CH3
CO2CH3
CH3O2C
CH2CO2CH3

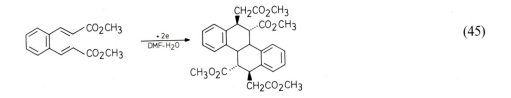

(45)

An intramolecularly cyclized product is obtained by the reduction of abscisic acid methyl ester [78].

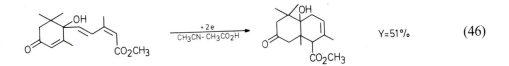

OH
O
CO2CH3
+2e
CH3CN-CH3CO2H
OH
O
CO2CH3
Y=51%

(46)

Intermolecular mixed hydrodimerization between conjugated olefins and α,β-unsaturated esters and ketones may also be achieved in some cases as shown by Eq. (47) [79].

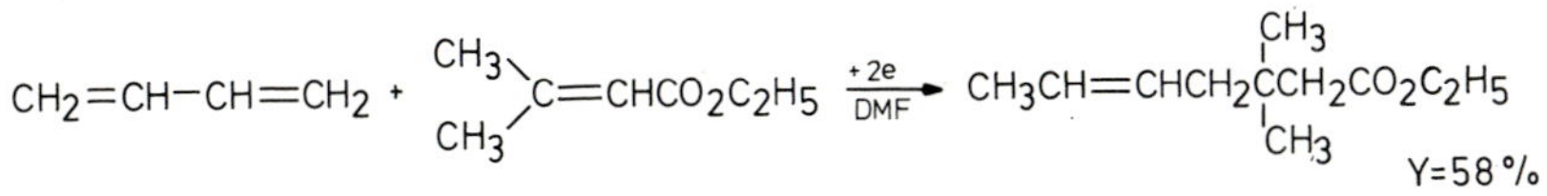

CH2=CH–CH=CH2 +
CH3
CH3
C=CHCO2C2H5
+2e
DMF
CH3CH=CHCH2CCH2CO2C2H5
CH3
CH3
Y=58%

(47)

3. Cathodic Reductions

When the pH is adequately controlled, N-ethylmaleimide is converted into the corresponding hydrodimer in aqueous solution [80]. The hydrodimerization carried out in the presence of alkylating agents gives the corresponding alkylated hydrodimers [81].

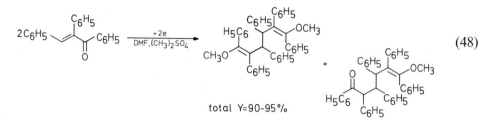

(48)

total Y=90-95%

Further reductive couplings are compiled in Table 7.

Table 7. Miscellaneous Reductive Coupling

Substrate	Product	Yield (%)	Ref.
C_6H_5 ... C_6H_5 ... C_6H_5 Br^-	C_6H_5 ... C_6H_5 C_6H_5 ... C_6H_5 C_6H_5 C_6H_5	61	[82]
1-Octyl bromide Fe(acac)$_3$	Hexadecane Octene Octane	93.9	[83]
$C_6H_5CH_2Cl$ Ni(acac)$_2$ $(C_6H_5)_3P$	Bibenzyl	87	[83]
$C_6H_5COCH=CHCl$	$C_6H_5COCH_2CH=CHCH_2COC_6H_5$ $C_6H_5COCH=CH-CH=CHCOC_6H_5$	40	[84]
C_4H_9N ... Cl^-	C_4H_9-N ... $N-C_4H_9$	45	[85]
CH_3S ... S ... SC_2H_5 CH_3S ... S BF_4^-	CH_3S ... S ... SC_2H_5 S ... SCH_3 CH_3S ... S ... S C_2H_5S ... S ... SCH_3	99.5	[86]
$(CH_3)_3SiCl$	$(CH_3)_3Si-Si(CH_3)_3$	95 (Current yield)	[87]
CH_3—⟨ ⟩—SCl	CH_3—⟨ ⟩—SS—⟨ ⟩—CH_3	85	[88]

References

1. a) Klemm, L. H., Iversen, P. E., Lund, H.: Acta Chem. Scand. *B28*, 593 (1974)
   b) Christensen, L., Iversen, P. E.: Acta Chem. Scand. *B33*, 352 (1979)
2. Iversen, P. E., Lund, H.: Acta Chem. Scand. *B28*, 827 (1974)
3. Curphey, T. J., Trivedi, L. D., Layloff, T.: J. Org. Chem. *39*, 3831 (1974) See also Hall, E. A. H., Moss, G. P., Utley, J. H. P., Weedon, B. C. L.: J. Chem. Soc., Chem. Commun. *1976*, 586
4. Gottlieb, R., Pfleiderer, W.: Justus Liebigs Ann. Chem. *1981*, 1451
5. Lund, H.: Acta Chem. Scand. *B31*, 424 (1977)
6. Shono, T., Nishiguchi, I., Ohmizu, H.: J. Am. Chem. Soc. *99*, 7396 (1977)
7. Lund, H., Degrand, C.: Tetrahedron Lett. *1977*, 3593
8. Shono, T., Nishiguchi, I., Ohmizu, H.: Chem. Lett. *1977*, 1021
9. Engels, R., Schäfer, H. J.: Angew. Chem. Int. Ed. Engl. *17*, 460 (1978)
10. Lund, H., Degrand, C.: Acta Chem. Scand. *B33*, 57 (1979)
11. Karrenbrock, F., Schäfer, H. J.: Tetrahedron Lett. *1978*, 1521
12. Shono, T., Ohmizu, H., Kawakami, S., Nakano, S., Kise, N.: Tetrahedron Lett. *22*, 871 (1981)
13. Shono, T., Kise, N., Yamazaki, A., Ohmizu, H.: Tetrahedron Lett. *23*, 1609 (1982)
14. Shono, T., Ohmizu, H., Kise, N.: Tetrahedron Lett. *23*, 4801 (1982)
15. Shono, T., Kise, N., Suzumoto, T.: J. Am. Chem. Soc. *106*, 259 (1984)
16. Karrenbrock, F., Schäfer, H. J., Langer, I.: Tetrahedron Lett. *1979*, 2915
17. Shono, T., Ohmizu, H., Kawakami, S., Sugiyami, H.: Tetrahedron Lett. *21*, 5029 (1980)
18. Shono, T., Matsumura, Y., Kashimura, S.: The 41st Annual Meeting of the Chemical Society of Japan (1980), Abstract, p. 671
19. Shono, T., Ohmizu, H.: Presented at the 3rd EUCHEM Meeting on Organic Electrochemistry, Pitlochry, Scotland, 1977
20. Abbot, E. M., Bellamy, A. J.: J. Chem. Soc., Perkin Trans. (2), *1978*, 254
21. Barba, F., Velasco, M. D., Guirado, A.: Synthesis *1981*, 625
22. Weinberg, N. L., Hoffmann, A. K., Reddy, R. B.: Tetrahedron Lett. *1971*, 2271
23. Tyssee, D. A., Baizer, M. M.: J. Org. Chem. *39*, 2819 (1974)
24. Tyssee, D. A., Baizer, M. M.: J. Org. Chem. *39*, 2823 (1974)
25. Wawzonek, S., Shradel, J. M.: J. Electrochem. Soc. *126*, 401 (1979)
26. Fuchs, P., Hess, U., Holst, H. H., Lund, H.: Acta Chem. Scand. *B35*, 185 (1981)
27. Root, D. K., Smith, W. H.: J. Electrochem. Soc. *129*, 1231 (1982)
28. See also a) Wawzonek, S., Blaha, E. W., Berkey, R., Runner, M. E.: J. Electrochem. Soc. *102*, 235 (1955)
   b) Wawzonek, S., Wearing, D.: J. Am. Chem. Soc. *81*, 2067 (1959)
   c) Wawzonek, S., Gundessen, A.: J. Electrochem. Soc. *107*, 537 (1960)
   d) Wawzonek, S., Duty, R. C., Wagenknecht, J. H.: J. Electrochem. Soc. *111*, 74 (1964)
   e) Wawzonek, S., Gundersen, A.: J. Electrochem. Soc. *111*, 324 (1964)
   f) Dietz, R., Peover, M. E.: Discuss. Faraday Soc. *45*, 154 (1968)
29. Shono, T., Mitani, M.: Nippon Kagaku Kaishi *1973*, 975
30. Shono, T., Mitani, M.: J. Am. Chem. Soc. *93*, 5284 (1971)
31. Shono, T., Nishiguchi, I., Ohmizu, H., Mitani, M.: J. Am. Chem. Soc. *100*, 545 (1978)

32. Shono, T., Nishiguchi, I., Ohmizu, H.: Chem. Lett. *1976*, 1233
33. Shono, T., Ohmizu, H., Kawakami, S.: Tetrahedron Lett. *1979*, 4091
34. Satoh, S., Taguchi, T., Itoh, M., Tokuda, M.: Bull. Chem. Soc. Jpn. *52*, 951 (1979)
35. Satoh, S., Suginome, H., Tokuda, M.: Bull. Chem. Soc. Jpn. *54*, 3456 (1981)
36. Baizer, M. M., Chruma, J. L.: J. Org. Chem. *37*, 1951 (1972)
37. Moore, W. M., Peters, D. G.: Tetrahedron Lett. *1972*, 453
38. Berube, D., Caza, J., Kimmerle, F. M., Lessard, J.: Can. J. Chem. *53*, 3060 (1975)
39. Shono, T., Mitani, M.: Nippon Kagaku Kaishi *1972*, 2370
40. a) Gassman, P. G., Rasmy, O. M., Murdock, T. O., Saito, K.: J. Org. Chem. *46*, 5457 (1981)
    b) See also Scheffold, R., Dike, M., Dike, S., Herold, T., Walder, L.: J. Am. Chem. Soc. *102*, 3642 (1980)
41. Nugent, S. T., Baizer, M. M., Little, R. D.: Tetrahedron Lett. *23*, 1339 (1982)
42. Shono, T., Yoshida, K., Ando, K., Usui, Y., Hamaguchi, H.: Tetrahedron Lett. *1978*, 4819
43. Shono, T., Usui, Y., Mizutani, T., Hamaguchi, H.: Tetrahedron Lett. *21*, 3073 (1980)
44. Shono, T., Usui, Y., Hamaguchi, H.: Tetrahedron Lett. *21*, 1351 (1980)
45. Shono, T., Miyamoto, T., Mizukami, M., Hamguchi, H.: Tetrahedron Lett. *22*, 2385 (1981)
46. Troll, T., Baizer, M. M.: Electrochim. Acta *20*, 33 (1975)
47. Lund, H., Simonet, J.: Bull. Soc. Chim. Fr. *1973*, 1843
48. a) Degrand, C., Canpagnon, P. L., Belot, G., Jacquin, D.: J. Org. Chem. *45*, 1189 (1980)
    b) See also Degrand, C., Jacquin, D.: Tetrahedron Lett. *1978*, 4955
49. Hansen, P. E., Berg, A., Lund, H.: Acta Chem. Scand. *B30*, 267 (1976)
50. Kristensen, L. H., Lund, H.: Acta Chem. Scand. *B33*, 735 (1979)
51. Moore, W. M., Finkelstein, M., Ross, S. D.: Tetrahedron *36*, 727 (1980)
52. Barry, J. E., Finkelstein, M., Moore, W. M., Ross, S. D.: J. Org. Chem. *47*, 1292 (1982)
53. Iversen, P. E., Lund, H.: Acta Chem. Scand. *B28*, 827 (1974)
54. Wagenknecht, J. H.: J. Org. Chem. *42*, 1836 (1977)
55. Fry, A. J., Scoggins, R.: Tetrahedron Lett. *1972*, 4079
56. Elofson, R. M., Gadallah, F. F.: J. Org. Chem. *36*, 1769 (1971)
57. Gottlieb, R., Neumeyer, J. L.: J. Am. Chem. Soc. *98*, 7108 (1976)
58. a) Grimshaw, J., Grimshaw, J. T.: Tetrahedron Lett. *1974*, 993
    b) Grimshaw, J., Haslett, R. J., Grimshaw, J. T.: J. Chem. Soc., Perkin Trans. (1), *1977*, 2448
59. Begley, W. J., Grimshaw, J., Grimshaw, J. T.: J. Chem. Soc., Perkin Trans. (1), *1974*, 2633
60. Grimshaw, J., Haslett, R. J.: J. Chem. Soc., Perkin Trans. (1), *1980*, 657
61. a) Feoktistov, L. G., Lund, H.: Organic Electrochemistry (ed.) Baizer, M. M., p. 347, Dekker, New York 1973
    b) Beck, F.: Angew. Chem. Int. Ed. Engl. *11*, 760 (1972)
62. a) Curphey, T. J., Amelotti, C. W., Layloff, T. P., McCartney, R. L., Williams, J. H.: J. Am. Chem. Soc. *91*, 2817 (1969)
    b) Curphey, T. J., McCartney, R. L.: Tetrahedron Lett. *1969*, 5295
    c) See also Kariv, E., Cohen, B. J., Gileadi, E.: Tetrahedron *27*, 805 (1972)
    d) Thomsen, A. D., Lund, H.: Acta Chem. Scand. *25*, 1576 (1971)

63. Armand, J., Boulares, L.: Can. J. Chem. *54*, 1197 (1976)
64. Sioda, R. E., Terem, B., Utley, J. H. P., Weedon, B. C. L.: J. Chem. Soc. Perkin Trans. (1), *1976*, 561
65. Mandell, L., Hamilton, H., Day, R. A., Jr.: J. Org. Chem. *45*, 1710 (1980)
66. Powell, L. A., Wightman, R. M.: J. Am. Chem. Soc. *101*, 4412 (1979)
67. van Tilborg, W. J. M., Smit, C. J.: Tetrahedron Lett. *1977*, 3651
68. Nonaka, T., Asai, M.: Bull. Chem. Soc. Jpn. *51*, 2976 (1978)
69. Touboul, E., Dana, G.: J. Org. Chem. *44*, 1397 (1979)
70. Rusling, J. F., Zuman, P.: J. Org. Chem. *46*, 1906 (1981)
71. Köster, K., Wendt, H.: J. Electroanal. Chem. *138*, 209 (1982)
72. a) Horner, L., Skaletz, D. H.: Tetrahedron Lett. *1970*, 1103
    b) Horner, L., Skaletz, D. H.: Justus Liebigs Ann. Chem. *1975*, 1210
73. Baizer, M. M., Petrovich, J. P.: Progress in Physical Organic Chemistry (ed.) Streitweiser, A., Taft, R. W., Vol. 7, p. 189, Interscience, New York 1970
74. a) Parker, V. D.: Acta Chem. Scand. *B35*, 147, 149, 279, 295 (1981); *B36*, 260 225 (1982)
    b) Lamy, E., Nadjo, L., Saveánt, J. M.: J. Electroanal. Chem. *50*, 141 (1974)
75. Baizer, M. M.: Organic Electrochemistry (ed.) Baizer, M. M., p. 679, Dekker, New York 1973
76. Mandell, L., Daley, R. F., Day, R. A., Jr.: J. Org. Chem. *41*, 4087 (1976)
77. a) Andersson, J., Eberson, L., Svensson, C.: Acta Chem. Scand. *B32*, 234 (1978)
    b) Andersson, J., Eberson, L., Svensson, C.: J. Chem. Soc., Chem. Commun. *1976*, 565
78. Terem, B., Utley, J. H. P.: Electrochim. Acta *24*, 1081 (1979)
79. Thomas, H. G., Thönnessen, F.: Chem. Ber. *112*, 2786 (1979)
80. Zoutendam, P. H., Kissinger, P. T.: J. Org. Chem. *44*, 758 (1979)
81. Troll, T., Elbe, W., Ollmann, G. W.: Tetrahedron Lett. *22*, 2961 (1981)
82. Shono, T., Toda, T., Oda, R.: Tetrahedron Lett. *1970*, 369
83. Jennings, P. W., Pillsburg, D. G., Hall, J. L., Brice, V. T.: J. Org. Chem. *41*, 719 (1976)
84. Matschiner, H., Voigtländer, R., Liesenberg, R.: Electrochim. Acta *24*, 331 (1979)
85. a) Gale, R. J., Osteryoung, A.: J. Electrochem. Soc. *127*, 2167 (1980);
    b) see also Rusling, J. F.: J. Electroanal. Chem. *125*, 447 (1981)
86. a) Moses, P. R., Chambers, J. Q.: J. Am. Chem. Soc. *96*, 945 (1974);
    b) see also Engler, E. M., Green, D. C.: J. Chem. Soc., Chem. Commun. *1976*, 148
87. Hengge, E., Litscher, G.: Angew. Chem. *88*, 414 (1974)
88. Johansson, B. L., Persson, B.: Acta Chem. Scand. *B32*, 431 (1978)

3.2 Cathodic Eliminations

In this section 1,1-, 1,2-, 1,3-, and 1,4-eliminations from 1,n (n = 1 ~ 4) disubstituted compounds are described. The simple reductive cleavage of a single bond is not mentioned except the reductive elimination of a protecting group, i.e. deprotection.

3. Cathodic Reductions

3.2.1 1,1-Elimination

It has been mentioned in the previous section that CCl_3^- is generated by the cathodic reduction of CCl_4 and easily adds to an activated double bond or a carbonyl group. When CCl_3^- is formed in the absence of electrophiles or proton donors, however, it decomposes into Cl^- and dichlorocarbene which can be trapped by olefins [1, 2].

$$CCl_4 \xrightarrow[CHCl_3]{+2e, -2Cl^-} [CCl_2:] \xrightarrow{R^1R^2C=CHR^3} R^1R^2C\!\!-\!\!CHR^3 \qquad (1)$$

$$\underset{CCl_2}{\diagdown\diagup}$$

$R^1, R^2, R^3 = CH_3;\ Y = 82\%$

$R^1 = CH_3, R^2 = C_6H_5, R^3 = H\ ;\ Y = 67\%$

Difluorocarbene is formed similarly by the reduction of CF_2Br_2 [3].

$$CF_2Br_2 \xrightarrow[CH_2Cl_2]{+2e, -2Br^-} [CF_2:] \xrightarrow{C_6H_5-\overset{CH_3}{\underset{}{C}}=CH_2} C_6H_5-\overset{CH_3}{\underset{CF_2}{\underset{\diagdown\diagup}{C}}}\!\!-\!\!CH_2 \qquad (2)$$

$Y = 57\%$

The formation of carbenes has also been observed in the reduction of geminal dihalonorbornanes [4].

$$\xrightarrow[DMF]{+2e} \qquad \longrightarrow \qquad \qquad (3)$$

$Y = 60\%$

The reduction of 1,1-diaryl-2,2-dihaloethylenes shows an interesting rearrangement in which one of the aryl groups migrates to the 2-position after the elimination of two geminal halogen atoms as halogen anions [5].

$$\underset{Ar}{\overset{Ar}{\diagdown}}C=C\underset{X}{\overset{X}{\diagup}} \xrightarrow[-X^-]{+2e} \underset{Ar}{\overset{Ar}{\diagdown}}C=C\underset{\_}{\overset{X}{\diagup}} \xrightarrow{-X^-} ArC\equiv CAr \qquad (4)$$

$Ar = p\text{-}CH_3O\text{-}C_6H_4 \quad Y = 82\%$

$X = Cl$

The formation of nitrene by the reduction of N,N-dichloro-p-toluene-sulfonamide has been proposed on the basis of the structure of the product [6].

$$CH_3-\!\!\left\langle\!\bigcirc\!\right\rangle\!\!-SO_2NCl_2 \xrightarrow[-2Cl^-]{+2e} [TsN:] \xrightarrow{} TsNH \qquad Y = 32\% \qquad (5)$$

Although the intervention of a carbene anion radical has been proposed in the reduction of diphenyldiazomethane [7] or 9-diazofluorene [8] in DMF, a mechanism involving formation of this active species has not been established so far [9].

Carbon monoxide is eliminated when 4,5-bis(alkylthio)-1,3-dithiol-2-ones are electrochemically reduced in the presence of alkylating agents [10].

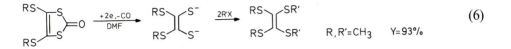

$$\tag{6}$$

3.2.2 1,2-Elimination

The synthesis of olefins by electroreductive 1,2-elimination of vicinal dihalides (Eq. (7)) is not necessarily useful in organic synthesis since the starting vicinal dihalides are in most cases prepared from olefins which are the same as those obtained by the 1,2-elimination.

$$R^1R^2C\!-\!CR^3R^4 \xrightarrow{+2e} R^1R^2C\!=\!CR^3R^4 \tag{7}$$
$$\overset{|}{X}\quad\overset{|}{X}$$

However, if the dihalides can be prepared by different ways, the electroreductive elimination is an attractive method for the synthesis of olefins since very mild reaction conditions are applied and a high stereospecificity is often achieved.

Since the electrochemical study on mechanism of the 1,2-elimination has already been surveyed in detail [11], only some typical reactions which are useful in organic synthesis are mentioned.

A high stereospecificity of the 1,2-elimination has been observed in the reduction of some vicinal dibromides [12, 13].

$$CH_3\!-\!CH\!-\!CH\!-\!CH_3 \xrightarrow[DMF]{+2e,-2Br^-} CH_3\,CH\!=\!CHCH_3 \tag{8}$$
$$\overset{|}{Br}\quad\overset{|}{Br}$$

meso   trans   Y=100%

dl   cis   Y=100%

This stereospecificity suggests the concerted exclusive *trans* elimination, if the appropriate conformation is realized in the transition state.

The electrochemical elimination of halogens is one of the most efficient methods for the preparation of highly strained cyclic olefins. Some typical examples are shown below.

Y=~100%   (9) [14]

3. Cathodic Reductions

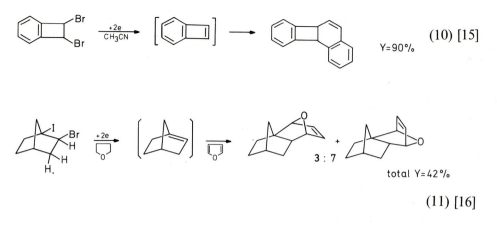

$$Y=90\% \qquad (10)\,[15]$$

$$3:7 \qquad \text{total } Y=42\%$$

$$(11)\,[16]$$

This elimination has been applied to the synthesis of an amino acid derivative which exhibits antibacterial activity [17].

$$Cl_3CCHCH\overset{NHCO_2CH_2C_6H_5}{\underset{CO_2CH_3}{}} \quad \underset{\substack{Cl}{}}{} \xrightarrow[CH_3OH, H_2O, HCl]{+2e} Cl_2C=CHCH\overset{NHCO_2CH_2C_6H_5}{\underset{CO_2CH_3}{}} \quad Y=92\% \qquad (12)$$

As Table 1 reveals vicinal disubstituted compounds other than dihalides also yield the corresponding unsaturated products through electroreduction.

A transformation of an ester group to a vinyl group [24] is an example of an application of the electroreductive 1,2-elimination to organic synthesis

$$RCO_2CH_3 + ArSO_2CH_2MgI \longrightarrow R\overset{O}{\overset{\|}{C}}CH_2SO_2Ar \xrightarrow{NaBH_4} R\overset{OH}{\overset{|}{C}H}CH_2SO_2Ar \xrightarrow{+2e} RCH=CH_2$$

$$R=C_{17}H_{35} \qquad Y=82\% \text{ (elimination)}$$

$$R= \qquad Y=76\%$$

$$R= HO(CH_2)_{10}CH_2- \qquad Y=80\% \qquad (13)$$

Elongation of carbonyl compounds has also effectively been achieved by this reductive elimination [25].

$$\overset{R^1}{\underset{R^2}{}}C=O + \overset{C_6H_5S}{\underset{CH_3O}{}}CHLi \longrightarrow \overset{R^1}{\underset{R^2}{}}\overset{OH}{\overset{|}{C}}-\overset{SC_6H_5}{\underset{OCH_3}{\overset{|}{C}H}} \xrightarrow[DMF]{+2e} \overset{R^1}{\underset{R^2}{}}C=CHOCH_3$$

$$R^1=C_6H_5CH_2CH_2, R^2=H, Y=96\% \text{ (elimination)}$$

$$R^1=C_9H_{19}, R^2=H \qquad Y=98\%$$

$$R^1, R^2=-(CH_2)_5- \qquad Y=92\% \qquad (14)$$

146

Table 1. Elimination of Some Vicinal Disubstituted Compounds

Substrate	Reaction conditions	Product	Yield (%)	Ref.
C_6H_5\C=C/C_6H_5, AcO / \OAc	$-1.5\,V\;(Ag/Ag^+)$, DMF	$C_6H_5C{\equiv}CC_6H_5$	95	[18]
$C_6H_5CH{-}CHC_6H_5$, $\overset{|}{SC_6H_5}\;\overset{|}{OAc}$	$-1.7\,V\;(Ag/Ag^+)$, DMF	$C_6H_5CH{=}CHC_6H_5$	trans 92	[19]
$C_6H_5{-}CH{-}CH_2$, $\overset{|}{OH}\;\overset{|}{SO_2C_6H_4CH_3}$	$-1.75\,V\;(Ag/Ag^+)$, DMF C_6H_5OH	$C_6H_5CH{=}CH_2$	76	[20]
$C_6H_5CH{-}CHSO_2C_6H_5$, $\overset{|}{OAc}\;\overset{|}{C_6H_5}$ (threo)	$-1.45\,V\;(Ag/Ag^+)$, DMF AcOH	$C_6H_5CH{=}CHC_6H_5$	trans 90	[20]
(pyridyl)$\overset{|}{CH}$(cyclohexyl), $\overset{O_2N}{}$ $\overset{|}{OAc}$	$-1.8\,V\;(SCE)$, DMF	(pyridyl)$-CH{=}$(cyclohexylidene)	89	[21]
$C_6H_5CHCCl_3$, $\overset{|}{OH}$	$-1.3\,V\;(SCE)$, DMF HCl	$C_6H_5CH{=}CCl_2$	95	[22]
(cyclohexyl)$\overset{CCl_3}{\underset{OH}{}}$	$-1.5\,V\;(SCE)$, DMF HCl	(cyclohexylidene)${=}CCl_2$	70	[22]
(spiro cyclobutane)$\overset{OH}{\underset{SC_6H_5}{}}$	$-1.8\sim-2.2\,V\;(SCE)$, DMF	(spiro cyclobutane)${=}CH_2$	70	[23]
$CH_3(CH_2)_8\overset{OH}{\underset{CH_3}{C}}CH_2SC_6H_5$	$-1.8\sim-2.2\,V\;(SCE)$, DMF	$CH_3(CH_2)_8\overset{}{\underset{CH_3}{C}}{=}CH_2$	96	[23]

3.2.3 1,3- and 1,4-Elimination

Although it is not always a true elimination but similar to substitution with respect to its mechanism, the formation of three- and four-membered cyclic compounds from 1,3- and 1,4-disubstituted compounds is designated as elimination in this section.

The formation of cyclopropanes by reduction of 1,3-dihalides has been extensively studied [26], and a stepwise mechanism has been proposed on the basis of the nonstereospecificity of the reaction [27].

$$R^1CHCH_2CHR^2 \xrightarrow[-X^-]{+2e} R^1\bar{C}HCH_2CHR^2 \xrightarrow[-X^-]{} R^1CH{-}CHR^2 \overset{CH_2}{\triangle} \qquad (15)$$
$$\overset{|}{X}\quad\overset{|}{X}\qquad\qquad\qquad\overset{|}{X}$$

147

3. Cathodic Reductions

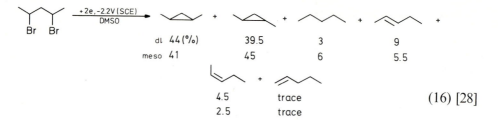

$$dl \ 44 \ (\%) \qquad 39.5 \qquad 3 \qquad 9$$
$$meso \ 41 \qquad 45 \qquad 6 \qquad 5.5$$

$$4.5 \qquad trace$$
$$2.5 \qquad trace$$

(16) [28]

It has been shown that the reaction mechanism greatly depends on the cathode potential, and a more negative potential results in a higher yield of the cyclic product [29], although the mechanism is also controlled by the structure of dihalides.

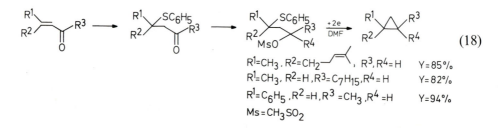

(17)

The synthesis of cyclopropanes through electroreductive 1,3-elimination has been applied to the transformation of α,β-unsaturated carbonyl compounds to cyclopropanes [30].

$R^1 = CH_3, R^2 = CH_2 -\!\!\!\!<, R^3, R^4 = H \qquad Y = 85\%$
$R^1 = CH_3, R^2 = H, R^3 = C_7H_{15}, R^4 = H \qquad Y = 82\%$
$R^1 = C_6H_5, R^2 = H, R^3 = CH_3, R^4 = H \qquad Y = 94\%$
$Ms = CH_3SO_2$

(18)

A variety of 1,3-dicarbonyl compounds has been converted to cyclopropanes by 1,3-elimination [31].

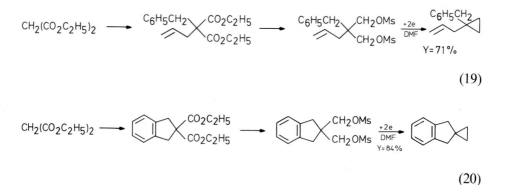

$Y = 71\%$

(19)

$Y = 84\%$

(20)

148

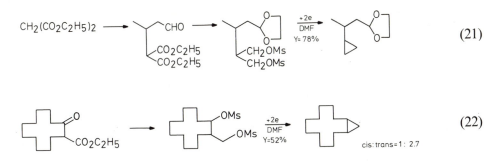

$$\text{(21)}$$

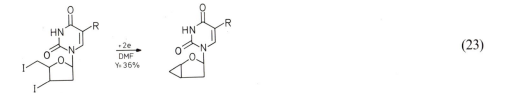

$$\text{(22)}$$

A carbohydrate derivative has also been prepared by 1,3-elimination [32].

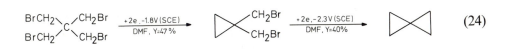

$$\text{(23)}$$

The cathodic 1,n-elimination is one of the most efficient methods for the synthesis of strained cyclic compounds. For example, spiropentane has been prepared by reductive elimination [33].

$$\text{(24)}$$

The intermediate formation of a propellane has been suggested in the reduction of 1,4-diiodonorbornane at low temperature [34].

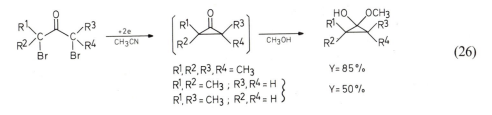

$$\text{(25)}$$

Cyclopropanones, especially cyclopropanone itself, are highly unstable whereas dimethyl- and tetramethylcyclopropanones have been synthesized readily by electroreduction of the corresponding α,α'-dihaloketones [35].

$$\text{(26)}$$

3. Cathodic Reductions

The acetal of cyclopropanone is synthesized from 1,3-dichloroacetone acetal.

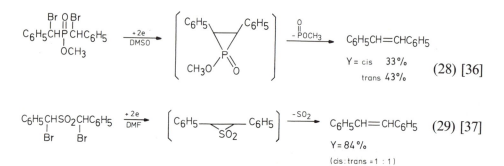

The intermediate formation of three-membered rings containing hetero-atoms such as sulfur and phosphorous has been suggested in the reduction of the corresponding 1,3-dihalides, although the heteroatoms are usually not contained in the stable final products.

The formation of a four-membered ring through 1,4-elimination has been observed in the reduction of $\alpha,\alpha,\alpha',\alpha'$-tetrabromo-o-xylene [38].

The further reduction of 1,2-dibromobenzocyclobutene has been mentioned previously (Section 3.2.2, Eq. (10)).

3.2.4 Deprotection

The reactions discussed in this section involve the removal of protecting groups by the electroreductive method which is usually much milder and more selective than the usual chemical methods. The electrochemical deprotection has been reviewed [39]. It can be roughly classified into two categories, (a) simple direct cleavage of ester, amide, or acetal bonds and (b) depro-tection through multiple cleavage of bonds. Deprotections of first type include the removal of tosyl, benzoyl, trityl, benzhydryl, benzyl, phenyl, cinnamyl, benzoyloxycarbonyl, and benzylidene groups, although the removal of these groups is not always possible by electroreduction. Since many examples concerning deprotections of this kind have already been reported [39], some deprotections of the other type are discussed in this section.

150

The addition of bromine to a double bond and subsequent electroreductive 1,2-elimination of bromine to regenerate the starting olefin can be used as the methods of protection and deprotection of a carbon—carbon double bond. Furthermore, the cathode potential necessary for the 1,2-elimination highly depends on the structure of the starting olefin. Thus, it is possible to protect selectively one double bond in a polyene. For example, in 4-vinyl-cyclohexene the less alkylated double bond is selectively protected [40].

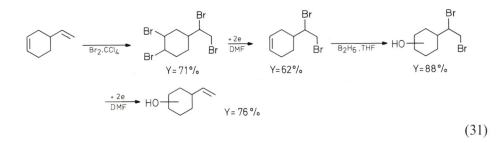

(31)

Hydroboration of 4-vinylcyclohexene usually takes place at the vinyl group.

When the bromination of a double bond is carried out with pyridinium hydrobromide perbromide (PyrHBr$_3$) and the subsequent electroreductive 1,2-elimination of bromine is achieved at -1.4 V (vs. SCE), the stereoconfiguration of the starting olefin is retained during protection and deprotection [41].

Selective protection of primary and secondary hydroxy groups may be performed by using a combination of two types of protecting agents and the electroreductive deprotection method [42].

Utilizing the large difference of the reduction potential between 9,10-dihydro-10-oxo-9-phenyl-9-anthracenyl ether and 4-cyanobenzyl ether, both primary and secondary hydroxy groups can be protected selectively.

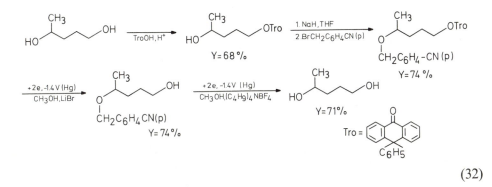

(32)

3. Cathodic Reductions

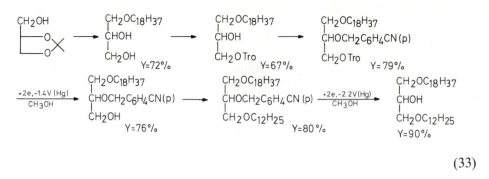

$$(33)$$

2,2,2-Trichloroethoxycarbonyl group may be used for the protection of hydroxy and amino groups since it is easily deprotected by reduction with zinc metal [43]. The electroreductive method has been found to be one of the most efficient methods for the elimination of the 2,2,2-trichloroethoxy-carbonyl moiety [44].

$$C_6H_5CH_2OCO_2CH_2CCl_3 \xrightarrow[CH_3OH]{+2e,-1.5V(SCE)} C_6H_5CH_2OH \qquad Y=70\% \tag{34}$$

$$C_6H_5CH_2SCO_2CH_2CCl_3 \xrightarrow[CH_3OH]{+2e,-1.5V(SCE)} C_6H_5CH_2SH \qquad Y=90\% \tag{35}$$

$$\underset{\underset{CO_2H}{|}}{CH_3CONHCHCH_2SCO_2CH_2CCl_3} \xrightarrow[CH_3OH]{+2e,-1.6V(SCE)} \underset{\underset{CO_2H}{|}}{CH_3CONHCHCH_2SH} \qquad Y=100\% \tag{36}$$

The electrodeprotection proceeds as follows.

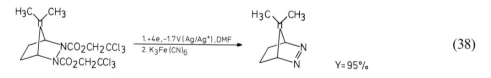

$$(37)$$

This method has been applied to the formation of bicyclic azo compounds [45].

$$(38) \qquad Y=95\%$$

References

1. Wawzonek, S., Duty, R.: J. Electrochem. Soc. *108*, 1135 (1961)
2. Fritz, H. P., Kornrumpf, W.: Justus Liebigs Ann. Chem. *1978*, 1416
3. Fritz, H. P., Kornrumpf, W.: J. Electroanal. Chem. *100*, 217 (1979)

152

4. Fry, A. J., Reed, R. G.: J. Am. Chem. Soc. *94*, 8475 (1972)
5. Merz, A., Thumm, G.: Justus Liebigs Ann. Chem. *1978*, 1526
6. Fuchigami, T., Nonaka, T., Iwata, K.: J. Chem. Soc., Chem. Commun. *1976*, 951
7. McDonald, R. N., January, J. R., Borhani, K. J., Hawley, M. D.: J. Am. Chem. Soc. *99*, 1268 (1977)
8. McDonald, R. N., Borhani, K. J., Hawley, M. D.: J. Am. Chem. Soc. *100*, 995 (1978)
9. a) Bethell, D., Galsworthy, P. J., Handoo, K. L., Parker, V. D.: J. Chem. Soc., Chem. Commun. *1980*, 534
   b) Parker, V. D., Bethell, D.: Acta Chem. Scand. *B34*, 617 (1980)
10. Falsig, M., Lund, H.: Acta Chem. Scand. *B34*, 591 (1980)
11. Rifi, M. R.: Organic Electrochemistry (ed.) Baizer, M. M., p. 279, Dekker, New York 1973
12. Casanova, J., Rogers, H. R.: J. Org. Chem. *39*, 2408 (1974)
13. Lund, H., Hobolth, E.: Acta Chem. Scand. *B30*, 895 (1976)
14. a) Casanova, J., Rogers, H. R.: J. Org. Chem. *39*, 3803 (1974)
   b) Wiberg, K. B., Bailey, W. F., Jason, M. E.: J. Org. Chem. *39*, 3803 (1974)
15. Rieke, R. D., Hudnall, P. M.: J. Am. Chem. Soc. *95*, 2646 (1973)
16. Stamm, E., Walder, L., Keese, R.: Helv. Chim. Acta *61*, 1545 (1978)
17. a) Urabe, Y., Iwasaki, T., Matsumoto, K., Miyoshi, M.: Tetrahedron Lett. *1975*, 997
   b) Iwasaki, T., Urabe, Y., Ozaki, Y., Miyoshi, M., Matsumoto, K.: J. Chem. Soc., Perkin Trans. (1), *1976*, 1019
18. Martigny, P., Michel, M. A., Simonet, J.: J. Electroanal. Chem. *73*, 373 (1976)
19. Martigny, P., Simonet, J.: J. Electroanal. Chem. *81*, 407 (1977)
20. Gambino, S., Martigny, P., Mousset, G., Simonet, J.: J. Electroanal. Chem. *90*, 105 (1978)
21. Petsom, A., Lund, H.: Acta Chem. Scand. *B34*, 614 (1980)
22. Merz, A.: Angew. Chem. Int. Ed. Engl. *16*, 57 (1977)
23. Shono, T., Matsumura, Y., Kashimura, S., Kyutoku, H.: Tetrahedron Lett. *1978*, 2807
24. Shono, T., Matsumura, Y., Kashimura, S.: Chem. Lett. *1978*, 69
25. Shono, T., Matsumura, Y., Kashimura, S.: Tetrahedron Lett. *21*, 1545 (1980)
26. a) Rifi, M. R.: Collect. Czech. Chem. Commun. *36*, 932 (1971)
   b) Rifi, M. R.: Tetrahedron Lett. *1969*, 1043
   c) Rifi, M. R.: J. Am. Chem. Soc. *89*, 4442 (1967)
27. Fry, A. J., Britton, W. E.: Tetrahedron Lett. *1971*, 4363
28. Fry, A. J., Britton, W. E.: J. Org. Chem. *38*, 4016 (1973)
29. Wiberg, K. B., Epling, G. A.: Tetrahedron Lett. *1974*, 1119
30. Shono, T., Matsumura, Y., Kashimura, S., Kyutoku, H.: Tetrahedron Lett. *1978*, 1205
31. Shono, T., Matsumura, Y., Tsubata, K., Sugihara, Y.: J. Org. Chem. *47*, 3090 (1982)
32. Adachi, T., Iwasaki, T., Miyoshi, M., Inoue, I.: J. Chem. Soc., Chem. Commun. *1977*, 248
33. Rifi, M. R.: J. Org. Chem. *36*, 2017 (1971)
34. a) Carroll, W. F., Jr., Peters, D. G.: Tetrahedron Lett. *1978*, 3543
   b) Carroll, W. F., Jr., Peters, D. G.: J. Am. Chem. Soc. *102*, 4127 (1980)
   c) see also Wiberg, K. B., Bailey, W. F., Jason, M. E.: J. Org. Chem. *41*, 2711 (1976)

35. a) van Tilborg, W. J. M., Plomp, R., de Ruiter, R., Smit, C. J.: Recueil, J. Roy. Netherl. Chem. Soc. *99*, 206 (1980);
       see also b) Fry, A. J., Hong, S. S. S.: J. Org. Chem. *46*, 1962 (1981)
    c) Fry, A. J., O'Dea, J. J.: J. Org. Chem. *40*, 3625 (1975),
    d) Dirlam, J. P., Eberson, L., Casanova, J.: J. Am. Chem. Soc. *94*, 240 (1972)
36. Fry, A. J., Chung, L. L.: Tetrahedron Lett. *1976*, 645
37. Fry, A. J., Ankner, K., Handa, V.: Tetrahedron Lett. *22*, 1791 (1981)
38. Rampazzo, L., Inesi, A., Bettolo, R. M.: J. Electroanal. Chem. *83*, 341 (1977)
39. Mairanovsky, V. G.: Angew. Chem. Int. Ed. Engl. *15*, 281 (1976)
40. Husstedt, U., Schäfer, H. J.: Synthesis *1979*, 964
41. Husstedt, U., Schäfer, H. J.: Tetrahedron Lett. *22*, 623 (1981)
42. Stouwe, C. v. d., Schäfer, H. J.: Chem. Ber. *114*, 946 (1981)
43. a) Windholz, T. B., Johnson, D. B. R.: Tetrahedron Lett. *1967*, 2555
    b) Karaday, S., Pines, S. H., Weinstock, L. M., Roberts, F. E., Brenner, G. S., Hoinowski, A. M., Chang, T. Y., Sletzinger, M.: J. Am. Chem. Soc. *94*, 1411 (1972)
    c) Rakhit, S., Bagli, J. F., Deghenghi, R.: Can. J. Chem. *47*, 2906 (1969)
44. a) Semmelhack, M. S., Heinsohn, G. E.: J. Am. Chem. Soc. *94*, 5139 (1972);
       see also b) Grimshaw, J.: J. Chem. Soc. *1965*, 7136,
    c) Cook, A. F.: J. Org. Chem. *33*, 3589 (1968)
45. Little, R. D., Carroll, G. L.: J. Org. Chem. *44*, 4720 (1979)

3.3 Cathodic Generation of Active Bases

Although numerous organic reactions using bases as catalysts have been studied, the types of bases available are rather limited. Hence, the generation of novel active bases *in situ* with adequate control of the amount of the base is often highly desirable in the development of new synthetic reactions. Cathodic reduction is evidently one of the most efficient methods to generate such bases in the reaction systems. A variety of reactions using this cathodic method have been studied. They may be classified into two categories, namely, (a) reactions catalyzed by the electrogenerated bases (EGB), and (b) nucleophilic addition of the bases to electrophiles. In the case of the reaction belonging to the second category the role of an EGB is not a base catalyst but a nucleophile. Some of these reactions have already been discussed in Sections 3.1.1 and 3.1.2.

The reactions assigned to the first category may be further classified into two groups. In the first group, the chemical behavior of the EGB is clearly understood, whereas the mechanism is still ambiguous in the second group though the reaction clearly involves some base catalyst. In this section, the reactions belonging to the first group of the first category are mainly mentioned.

3.3.1 Reactions of Electrogenerated Bases

One of the early studies in which the anionic species was used as a base is the reduction of cyanomethyltriphenylphosphonium bromide in which the cyanomethyl anion formed by the cathodic cleavage of the onium salt reacts as a base with the starting onium salt to yield the corresponding ylid [1].

$$(C_6H_5)_3\overset{+}{P}CH_2CN \xrightarrow{+2e} (C_6H_5)_3P + (CH_2CN)^-$$

$$(C_6H_5)_3\overset{+}{P}CH_2CN + (CH_2CN)^- \longrightarrow (C_6H_5)_3P=CHCN + CH_3CN \tag{1}$$

The formation of ylids by electroreduction of phosphonium salts has been confirmed by carrying out the reduction in the presence of carbonyl compounds [2, 3]. The yields of olefins are reasonable.

$$(C_6H_5)_3\overset{+}{P}CH_3 + \underset{C_2H_5}{\overset{C_4H_9}{>}}CHCHO \xrightarrow[CH_3CN,\,Al-Cathode]{+2e} \underset{C_2H_5}{\overset{C_4H_9}{>}}CHCH=CH_2 \tag{2}$$
$$I^- \qquad\qquad\qquad\qquad\qquad\qquad\qquad Y=89\%$$

$$(C_6H_5)_3\overset{+}{P}C_4H_9 + (C_2H_5)_2CHCHO \xrightarrow[CH_3CN,\,Al-Cathode]{+2e} (C_2H_5)_2CHCH=CHC_3H_7 \tag{3}$$
$$I^- \qquad\qquad\qquad\qquad\qquad\qquad cis-isomer$$
$$\qquad\qquad\qquad\qquad\qquad\qquad\qquad Y=82\%$$

$$(C_6H_5)_3\overset{+}{P}(CH_2)_3CO_2CH_3 + C_9H_{19}CHO \xrightarrow[CH_3CN,\,Al-Cathode]{+2e} C_9H_{19}CH=CH(CH_2)_2CO_2CH_3 \tag{4}$$
$$I^- \qquad\qquad\qquad\qquad\qquad\qquad\qquad cis-isomer$$
$$\qquad\qquad\qquad\qquad\qquad\qquad\qquad\qquad Y=69\%$$

One of the possible mechanisms of the formation of ylids involves abstraction of a proton from the starting phosphonium salt by the base formed by electroreductive cleavage of the onium salt [4, 5], though the mechanism in which ylids are formed by direct one-electron reduction of phosphonium salts or through abstraction of a proton from the phosphonium salt by a base generated from a solvent cannot always be denied.

Formation of the anionic species by cathodic reduction of a sulfonium salt has also been suggested in the formation of 4-phenylbutyronitrile from benzyldimethylsulfonium p-toluenesulfonate and acrylonitrile, though this process is not a base-catalyzed reaction but an addition of the anionic species to the active olefin [6].

$$(CH_3)_2\overset{+}{S}CH_2C_6H_5 + CH_2=CHCN \xrightarrow{+2e} C_6H_5CH_2CH_2CH_2CN \tag{5}$$
$$^-OTs \qquad\qquad\qquad\qquad\qquad\qquad Y=29\%$$

155

3. Cathodic Reductions

Although the reaction mechanism is not clear as yet, sulfonium ylids are also formed by the reduction of sulfonium salts [7, 8].

$$(CH_3)_3S^+I^- + C_6H_5CHO \xrightarrow[DMSO]{+2e} C_6H_5CH\overset{\displaystyle\diagdown}{\underset{O}{\diagup}}CH \qquad Y = 32\% \tag{6}$$

An ammonium ylid has been suggested to be generated as an intermediate in the electrolytic Stevens rearrangement [9].

$$C_6H_5COCH_2\overset{+}{\underset{\underset{CH_2C_6H_5}{|}}{N}}(CH_3)_2 \xrightarrow[DMF]{+2e} \left[C_6H_5CO\overline{C}H\overset{+}{\underset{\underset{CH_2C_6H_5}{|}}{N}}(CH_3)_2 \right] + C_6H_5COCH_3 + C_6H_5CH_2N(CH_3)_2$$

$$\longrightarrow C_6H_5COCH\underset{\underset{CH_2C_6H_5}{|}}{N}(CH_3)_2 \qquad Y = 36\% \tag{7}$$

In contrast to the ambiguity concerning the mechanism of the formation of ylids by electroreduction of phosphonium and sulfonium salts, the formation of stilben by cathodic reduction of a solution of azobenzene, benzaldehyde and benzyltriphenylphosphonium bromide in DMF is clearly explained by the generation of an anionic species from azobenzene followed by abstraction of a proton from the onium salt [10].

$$C_6H_5N{=}NC_6H_5 + (C_6H_5)_3\overset{+}{P}CH_2C_6H_5 + C_6H_5CHO \xrightarrow[DMF]{+2e} C_6H_5NHNHC_6H_5 + C_6H_5CH{=}CHC_6H_5$$

$$Y = \sim 100\% \qquad\qquad Y = 98\% \tag{8}$$

Compounds which form EGB by electroreduction are called probases (PB), and azobenzene and its derivatives have been shown to be typical probases in the utilization of EGB in organic synthesis [11]. One of the most typical examples of using azobenzene as a probase is the carboxylation of phenylacetate to phenylmalonate [12].

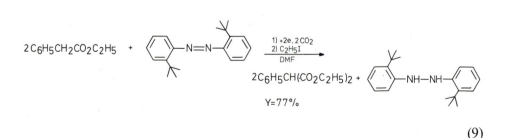

$$(9)$$

156

Besides azobenzene, active olefins such as ethyl acrylate and tetraethyl ethenetetracarboxylate are also effective probases.

$$CH_2(CO_2C_2H_5)_2 \quad + \quad CH_2{=}CHCO_2C_2H_5 \quad \xrightarrow[\text{DMF}]{+2e} \quad C_2H_5O_2CCH_2CH_2CH(CO_2C_2H_5)_2 \quad (10)\ [13]$$
$$Y=77\%$$

$$CH_2(CO_2C_2H_5)_2 \quad + \quad (C_2H_5O_2C)_2C{=}C(CO_2C_2H_5)_2 \quad \xrightarrow[\text{DMF}]{+2e} \quad (C_2H_5O_2C)_2\underset{\underset{\displaystyle CH(CO_2C_2H_5)_2}{|}}{C}{-}CH(CO_2C_2H_5)_2$$
$$Y=93\%$$

$$(11)\ [13]$$

The anionic species formed from the active olefin abstracts a proton from the active methylene compound. The resulting anion of the latter adds to the active olefin to give an anionic adduct which abstracts a proton from the active methylene compound to form the final product and to regenerate the anion of the active methylene compound. Thus, the EGB formed from an active olefin acts as a catalyst. Carboxylation of the intermediate anions may also occur [14].

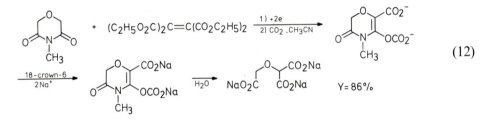

$$(12)$$

It has been observed that the cathodic reduction of acetonitrile in the presence of a suitable supporting electrolyte yields a cyanomethyl anion [15] which reacts with alkylboranes or the carbonyl group of a carbonyl compound, ester or amide, although the cyanomethyl anion in these reactions acts as a nucleophile rather than as a base.

$$C_6H_5COCH_3 \quad + \quad CH_3CN \quad \xrightarrow[\text{CH}_3\text{CN, (C}_2\text{H}_5)_4\text{NBF}_4]{+2e} \quad C_6H_5{-}\underset{\underset{\displaystyle CH_2CN}{|}}{\overset{\overset{\displaystyle OH}{|}}{C}}{-}CH_3 \quad \longrightarrow \quad \longrightarrow$$

$$\underset{\displaystyle Y=30\text{-}50\%}{C_6H_5\overset{\overset{\displaystyle CH_3}{|}}{C}HCH_2CN} \quad + \quad \underset{\displaystyle Y=35\text{-}45\%}{C_6H_5\underset{\underset{\displaystyle CH_2CN}{|}}{\overset{\overset{\displaystyle CH_3}{|}}{C}}{-}CH_2CN} \qquad\qquad (13)\ [16]$$

$$(C_6H_{13})_3B \quad + \quad CH_3CN \quad \xrightarrow[\text{CH}_3\text{CN, (C}_2\text{H}_5)_4\text{NI}]{} \quad 3C_7H_{15}CN \quad \begin{array}{l} Y=52\% \text{ based on borane} \\ \text{current } Y=16\% \end{array} \qquad (14)\ [17]$$

3. Cathodic Reductions

$$C_6H_5CO_2CH_3 \ + \ CH_3CN \ \xrightarrow[\text{CH}_3\text{CN, (C}_3\text{H}_7)_4\text{NClO}_4]{+2e} \ C_6H_5COCH_2CN \qquad Y=45\% \qquad (15) \ [18]$$

Base-catalyzed reactions using cyanomethyl anion as a base are not known.

The cathodic reduction of succinimide in acetonitrile or DMF [19] yields a succinimide anion which can be alkylated by alkyl halides or tetraalkylammonium salts. However, this anion is not a sufficiently strong base to catalyze base-catalyzed reactions.

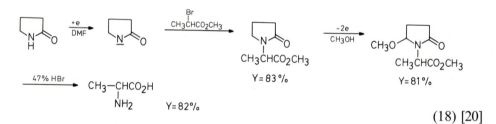

$$(17)$$

Pyrrolidone anion which can be generated in a similar way is a stronger base and nucleophile than the succinimide anion. The synthesis of an amino acid using pyrrolidone anion as the starting material is described below.

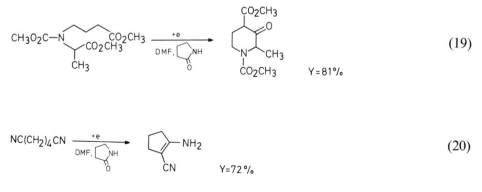

This pyrrolidone anion catalyzes the Dieckmann and the Thorpe reactions [21].

158

The reactivity of the pyrrolidone anion formed in this cathodic reduction is unique since the same anion generated by the reaction of pyrrolidone with sodium hydride is less effective to promote the Dieckmann reaction under the same reaction conditions.

The pyrrolidone anion is also an effective base catalyst to promote the condensation of chloroform with aldehydes to yield α-trichloromethyl-alkanols [22].

The anionic species formed from carbonyl compounds promote some base-catalyzed reactions such as aldol condensation with a high current efficiency [23].

$$CH_3CH_2CHO \xrightarrow[DMF]{+[e]} C_2H_5CH=C\begin{smallmatrix}CH_3 \\ CHO\end{smallmatrix} \quad \text{current} \quad Y=1 \times 10^4 \% \tag{21}$$

$$C_6H_5CHO + C_3H_7CHO \xrightarrow[DMF]{+[e]} C_6H_5CH=C\begin{smallmatrix}C_2H_5 \\ CHO\end{smallmatrix} \quad \text{current} \quad Y=9 \times 10^3 \% \tag{22}$$

The electrochemically induced rearrangement of S,S-diarylbenzene-1,2-dicarbothioates to the isomeric 3,3-bis(p-X-phenylthio)phthalides is an example of a reaction in which no external anion is necessary, and the amount of electricity required is less than 0.1 F/mol [24].

$$\tag{23}$$

Cathodic reduction of molecular oxygen is one of the most efficient methods of generating superoxide ion. The superoxide ion which is used as a reagent displays highly versatile properties. Thus, it may behave as a reducing agent, oxidizing agent, base, or nucleophile [25, 26]. An example in which superoxide ion clearly behaves as a base is shown below [27].

$$2O_2^{-\cdot} + NO_2\text{—}\langle \rangle\text{—}NH_2 \rightleftharpoons O_2 + HO_2^- + NO_2\text{—}\langle \rangle\text{—}NH^- \tag{24}$$

In the following two examples, superoxide ion behaves as an oxidizing agent in the former reaction while as a nucleophile and reducing agent in the latter.

$$CH_3\text{—}\langle \rangle\text{—}NO_2 + O_2^{-\cdot} \longrightarrow HO_2C\text{—}\langle \rangle\text{—}NO_2 \quad Y=57\% \tag{25} [28]$$

3. Cathodic Reductions

$$C_6H_5COCl \xrightarrow{O_2^{-\cdot}} C_6H_5CO_3\cdot \xrightarrow{O_2^{-\cdot}} C_6H_5CO_3^- \xrightarrow{C_6H_5CH=CHC_6H_5} C_6H_5CH\!\!-\!\!CHC_6H_5$$

$$Y=41\%$$

(26) [29]

Primary and secondary nitriles may also be transformed to the corresponding carbonyl compounds by using electrogenerated superoxide ion.

$$(C_6H_5)_2CHCN \xrightarrow[-1.0\,V\ vs.\ SCE]{+[e],\ O_2} (C_6H_5)_2CO \qquad Y=95\%$$

(27) [30]

References

1. Wagenknecht, J. H., Baizer, M. M.: J. Org. Chem. *31*, 3885 (1966)
2. Shono, T., Mitani, M.: J. Am. Chem. Soc. *90*, 2728 (1968)
3. Shono, T., Kashimura, S., Mizukami, M., Muramatsu, S.: The 47th Annual Meeting of the Chemical Society of Japan (1983), Abstract, Vol. II, p. 931
4. a) Pardini, V. L., Roulier, L., Utley, J. H. P., Webber, A.: J. Chem. Soc. Perkin Trans. (2), *1981*, 1520;
   b) see also Utley, J. H. P., Webber, A.: J. Chem. Soc., Perkin Trans. (1), *1980*, 1154
5. a) Savéant, J. M., Binh, S. K.: Bull. Soc. Chim. Fr. *1972*, 3549
   b) Sevéant, J. M., Binh, S. K.: Electrochim. Acta *20*, 21 (1975)
   c) Savéant, J. M., Binh, S. K.: J. Org. Chem. *42*, 1242 (1977)
6. Baizer, M. M.: J. Org. Chem. *31*, 3847 (1966)
7. a) Shono, T., Mitani, M.: Tetrahedron Lett. *1969*, 687
   b) Shono, T., Akazawa, T., Mitani, M.: Tetrahedron *29*, 817 (1973)
8. Beak, P., Sullivan, T. A.: J. Am. Chem. Soc. *104*, 4450 (1982) and references cited therein
9. Iversen, P. E.: Tetrahedron Lett. *1971*, 55
10. Iversen, P. E., Lund, H.: Tetrahedron Lett. *1969*, 3523
11. Troll, T., Baizer, M. M.: Electrochim. Acta *20*, 33 (1975)
12. Hallcher, R. C., Baizer, M. M.: Justus Liebigs Ann. Chem. *1977*, 737
13. Baizer, M. M., Chruma, J. L., White, D. A.: Tetrahedron Lett. *1973*, 5209
14. Hallcher, R. C., White, D. A., Baizer, M. M.: J. Electrochem. Soc. *126*, 404 (1979)
15. Becker, B. F., Fritz, H. P.: Justus Liebigs Ann. Chem. *1976*, 1015
    b) Bellamy, A. J.: J. Chem. Soc., Chem. Commun. *1975*, 944
    c) Bellamy, A. J., Howat, G., MacKirdy, I. S.: J. Chem. Soc., Perkin Trans. (2), *1978*, 786;
    d) see also van Tilborg, W. J. M., Smit, C. J., Scheele, J. J.: Tetrahedron Lett. *1977*, 2113
17. Takhashi, Y., Tokuda, M., Itoh, M., Suzuki, A.: Chem. Lett. *1975*, 523
18. Kristenbrügger, L., Mischke, P., Voß, J., Wiegand, G.: Justus Liebigs Ann. Chem. *1980*, 461
19. Moore, W. M., Finkelstein, M., Ross, S. D.: Tetrahedron *36*, 727 (1980)

20. Shono, T., Kashimura, S., Nogusa, H.: The 47th Annual Meeting of the Chemical Society of Japan (1983), Abstract, Vol. II, p. 932
21. Shono, T., Kashimura, S.: The 47th Annual Meeting of the Chemical Society of Japan (1983), Abstract, Vol. II, p. 927
22. Shono, T., Kashimura, S., Ishizaki, K. Ishige, O.: Chem. Lett. *1983*, 1311
23. Shono, T., Kashimura, S., Ishizaki, K.: Electrochim. Acta *29*, 603 (1984)
24. Praefcke, K., Weichsel, C., Falsig, M., Lund, H.: Acta Chem. Scand. *B34*, 403 (1980)
25. Raff, E. L.: Chem. Soc. Rev. *6*, 195 (1977)
26. Sawyer, D. T., Gibian, M. J., Morrison, M. M., Seo, E. T.: J. Am. Chem. Soc. *100*, 627 (1978)
27. Hussey, C. L., Haher, T. M., Achord, M.: J. Electrochem. Soc. *127*, 1484 (1980)
28. a) Sagae, H., Fujihira, M., Osa, T., Lund, H.: Chem. Lett. *1977*, 793;
    b) see also Sagae, H., Fujihira, M., Lund, H., Osa, T.: Heterocycles *13*, 321 (1979)
29. Nagano, T., Arakene, K., Hirobe, M.: Chem. Pharm. Bull. *28*, 3719 (1980); yield was obtained by using superoxide ion generated from KO_2
30. a) Sugawara, M., Baizer, M. M.: Tetrahedron Lett. *24*, 2223 (1983);
    b) see also Allen, P. M., Foote, C. S., Baizer, M. M.: Synth. Commun. *12*, 123 (1982);
    c) Monte, W. T., Baizer, M. M., Little, R. D.: J. Org. Chem. *48*, 803 (1983)

Appendix

It is not the main purpose of this monograph to show detailed experimental procedures of electroorganic reactions. They can easily be extracted from the experimental parts of the cited original papers. The majority of organic chemists may, however, be still unfamiliar with electroorganic chemistry. Therefore the outlines of operation of three types of reactions are mentioned in this appendix to show that electroorganic reactions are easily accomplished by organic chemists.

1. Constant Current Oxidation

Transformation of 3-Acetoxy-3-*p*-Menthene to 4-*p*-Menthen-3-one

Into an undivided cell shown in Fig. 1 is put a solution of 9.8 g of 3-acetoxy-3-*p*-menthene (50 mmol) and 1.5 g of tetraethylammonium *p*-toluenesulfonate (5 mmol) in 50 ml of acetic acid. A constant current (0.1 A) is passed through the cell with external water cooling. After 2.7–3.0 *F*/mol of electricity has passed, 100 ml of water is added to the reaction mixture. The aqueous solution is extracted with ether, and distillation of the ethereal solution gives the product in 90–97% yield. Bp. 77 °C/10 mm.

2. Controlled Potential Reduction

Synthesis of 8-Oxotetrahydropalmatine

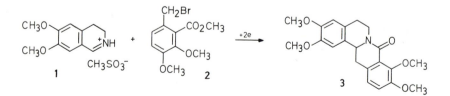

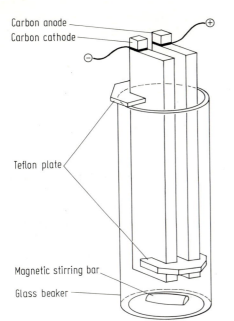

Carbon anode
Carbon cathode

Teflon plate

Magnetic stirring bar
Glass beaker

Fig. 1.

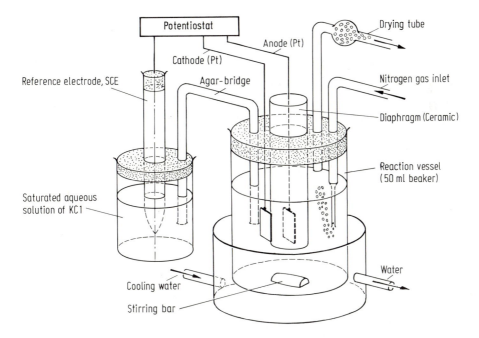

Potentiostat

Anode (Pt)

Cathode (Pt)

Drying tube

Reference electrode, SCE

Agar-bridge

Nitrogen gas inlet

Diaphragm (Ceramic)

Reaction vessel
(50 ml beaker)

Saturated aqueous
solution of KCl

Cooling water

Stirring bar

Water

Fig. 2.

163

The cell shown in Fig. 2 is one of the orthodox systems using an agar-bridge to connect a reference electrode with the reaction cell. There are a variety of methods for this connection, the orthodox cell shown in Fig. 2, however, is the most convenient from the standpoint of organic synthesis.

A solution of 0.526 g of 1 (2 mmol), 1.156 g of 2 (4 mmol), and 0.264 g of methanesulfonic acid (3 mmol) in 40 ml of DMF is placed in the cathodic chamber of the cell. The anodic chamber is filled with 5 ml of DMF containing 3 g of tetraethylammonium p-toluenesulfonate (10 mmol). The reaction is carried out at room temperature controlling the cathode potential at -1.8 V vs. SCE. After 2.5 F/mol of electricity has passed, solvent is removed in vacuo. The residue is poured into an aqueous solution of KOH, and is then extracted with $CHCl_3$. The chloroform solution is dried over $MgSO_4$, and the drying agent is filtered off. Evaporation of the solvent yields a residue, from which the product, 8-oxotetrahydropalmatine is isolated with column chromatography (silica gel, $CHCl_3/CH_3OH$). Yield is almost quantitative.

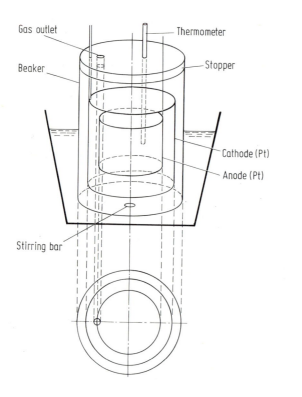

Gas outlet

Thermometer

Beaker

Stopper

Cathode (Pt)

Anode (Pt)

Stirring bar

Fig. 3.

3. Indirect Oxidation Using a Mediator

Oxidation of 2-Octanol to 2-Octanone

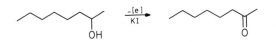

A mixture of 5.21 g of 2-octanol (40 mmol) and 1.66 g of KI (1 mmol) in 10 ml of water and 2 ml of *t*-butanol is placed in the cell shown in Fig. 3. A constant current of 0.2 A is passed through the cell with external cooling and vigorous stirring (the reaction mixture is not a homogeneous solution). After 10 F/mol of electricity has passed, the reaction mixture is extracted with ether. The product, 2-octanone is obtained by distillation of the ethereal solution. Yield is about 85%.

Subject Index